"Bird!"

An Exploration of Hawkwatching

Brian M. Wargo

BMW is an imprint of BMW Endeavors, LLC
78 Fawnvue Drive
McKees Rocks, PA 15136

Library of Congress Control Number 2016906135
Wargo, Brian M., 1971–
"Bird!": An Exploration of Hawkwatching

First Edition

ISBN: 978-1-945226-00-7 (paperback)
ISBN: 978-1-945226-01-4 (ebook)
ISBN: 978-1-945226-02-1 (audiobook)

ESSAYS

DEDICATION

This book is dedicated to all of the men and women who contribute
to our understanding of raptors.

INTRODUCTION
WHO SHOULD READ THIS BOOK

The cover of this book shows a man standing in the cold, waiting. But for what? Is the man thinking about a majestic Bald Eagle flying overhead? If so, how long will he have to wait before he sees one? What if the eagle does not appear? What will he do all day? Will he just stand there? Isn't that boring? These are questions that non-hawkwatchers contemplate when they hear about hawkwatching. Before I was a hawkwatcher, I too had similar questions.

The 22 self-contained essays that constitute this book, set out to explore why otherwise rational people stand outside in the elements and stare at the sky for hours on end. The goal is to expose the mind of the hawkwatcher and illuminate the culture of hawkwatching to the uninitiated. Because of this, interest in birding is not a requirement for enjoying this book. In fact, non-birders may enjoy it as much as an accomplished ornithologist. In other words, this book is for everyone, but no one in particular. The ideas gleaned from my experiences hawkwatching tap into the universality of naturalism; hawkwatching is simply the vehicle. The essays are, therefore, as general as they are personal.

The reader should not expect to become an expert hawk counter simply from reading this book. Although reading it may improve how you go about identifying hawks, help you understand what is involved with counting hawks, and provide awareness on how to act at a hawk site.

The order of the essays represents their chronological history; Essay One was written years ago and Essay Twenty-two just recently. However, they can be read in any order the reader chooses. Each essay will complement the others, portraying a different aspect of hawkwatching. The hope is that after reading each essay, the reader will have a deeper appreciation of hawkwatching, the hawkwatchers that partake in this odd activity, and the ecological world we share with the hawks.

This book, most of all, is a celebration of humanity, our relation to each other, and our avian cohabitants. So, please, just read, enjoy, and reflect. And, when at all possible, look up!

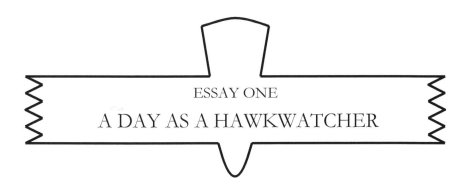

ESSAY ONE

A DAY AS A HAWKWATCHER

I wake at 4:30 a.m., naturally, without a clock. I am still alive—life is short; time to get up and live. I read, write, or study at this time every day. My mind is clear, sharp, focused. Today, I am reviewing my notes on raptors. This early morning plunge into the fine details of hawk plumage, shape, and behavior primes my mind for the upcoming day. I am counting hawks today, and I need to get into the zone. I pack, eat, and I am on the road by 6:30 a.m. for the hour drive to the hawk site.

I begin counting at 8:00 a.m., and the trek up to the site takes about 15 to 30 minutes, depending on how much I allow myself to sweat. On cold days, I hike slowly up the mountain wearing only a layer or two, taking breaks along the way, controlling my temperature. Today is cold, about 32 degrees; so, I am prepared with extra gear, which I am carrying instead of wearing.

I am beginning to sweat, so I slow down despite my desire to push on. My shoulders would like the journey to end sooner than later. I am carrying a spotting scope, the stand for the scope, a camera with a heavy 200-millimeter lens, and my coat, which is filled to capacity with gloves, a hat, a facemask, and an extra shirt or two that are

1

stuffed into the sleeves. My backpack is bulging at the seams, packed methodically to utilize every space available. It contains my binoculars, my bird books, a small stool, an umbrella, bear spray, my lunch, a thermos of hot tea, hand warmers, my Mountain Hardware shell pants, as well as a survival/emergency pack. Today, like most days, I am alone. Only the wind—the cold wind—will provide companionship.

The wind is my friend, and the wind is my foe. The raptors rely on the wind to provide lift. Their journey is long, and they must catch a free ride if they are to successfully migrate. Without wind, the raptor migration slows to a halt. For this reason, I welcome the wind. The trade-off is that the wind steals my heat. As I arrive at my post, a jettisoning rock called "Lovers Leap," the wind is absent and I feel warm. Too warm!

My steaming body attempts to dry in the cold, still air. As the water vaporizes, it takes my heat, which for now feels refreshing. I scan the sky; nothing flying yet. Mornings are often slow, but occasionally you catch the raptors as they are starting their day. These birds tend to be low, offering the closest views of the day. I unpack my daily log, check the weather on my mobile device, and record environmental data, including temperature, wind speed and direction, humidity, cloud cover, and atmospheric pressure. I will do this every hour for the rest of the day.

It is easy to write these values into the tabular data form, at least for now. Other thoughts stream through my head, and I write a few of the ideas down on a folded piece of paper that I carry with me at all times. My brain can be a firestorm of ideas, which are fleeting, only captured by writing them immediately. As I cool down, my thoughts slow, as well as the motivation to pull out the paper to write on. Before I know it, it is the end of the hour, and I must record the environmental data. No birds this hour, I write.

Looking up, I see the tail end of a bird streaking across the sky. It is red. In the time that I was writing, the first bird of the day was passing, and I nearly missed it. This is the problem with counting alone; birds pass even when you're not looking. I look intently at the sky, knowing that inattention equates to omitting migrating birds. For the next seven hours, I will peer into the deep blue, a prospect that some find absurd. As a young man, I would not only agree about the absurdity, but would have said it was impossible.

Having had what is modernly called attention deficit disorder, the prospect of focusing, standing, or doing the same thing for eight hours straight would have been preposterous. However, as a man grows, his abilities change. Being able to focus on the sky and the birds is Zen-like, which I liken to *becoming aware* of our place in the universe. For the next hour, I am intently aware. No birds, however, for this concerted effort.

Abruptly, around 10:00 a.m., the wind starts to blow from the south. The gentle, cool breeze removes the insulating air that has enveloped me up to this point. I begin the process of putting on additional layers. Within the next hour, I will put on another shirt, my shell pants, my coat, my facemask, my brimmed hat, and my gloves. I will use the hood on my coat to block as much of the wind as possible. My total layer value is now three, beginning with an Under Armour base layer, lined pants, and finally an outer wind-breaking, waterproof shell for my legs.

My torso is similarly outfitted with an Under Armour base layer, a synthetic shirt covered by a flannel shirt, and then my wind and waterproof coat. Winter hiking boots and insulated gloves, often with heat packs, rounds out my extremities. This helps take the chill off. But, I will still get cold. Being in shape, having low body fat, and being small in stature all conspire to deplete my heat. If I move around a bit, I warm up, but I must be careful not to move too much or carelessly, for I am on a cliff.

Several birds stream through; three Sharp-shinned Hawks and two more Red-tailed Hawks pass before an adult Bald Eagle appears in the sky. The majestic eagles are always a sight that generates excitement. Clearly, it is because of their size, their soaring ability, and their character. This is unfounded because Turkey Vultures are

3

similarly sized, master soarers, and seem to display more interest in people than the eagles. That is why I treat each raptor as equally valuable, for each plays an important part in the ecosystem, and should be granted equal affection.

I am lying! Nobody, including me, gets excited when a Turkey Vulture appears. I do, however, value the birds for the service they provide; but, I admit, I would rather see the more stoic, straight-flying, disinterested Bald Eagle.

Photo by Brian M. Wargo

An adult Bald Eagle provides an easy and unmistakable identification.

After marking the raptors on my log sheet, I recognize that it is lunchtime. I have been hungry since hiking up the hill, and I finally concede to my growling stomach. I pull out the small tri-leg stool for lunch and sit down for the first time of the day. To eat, I need to remove my gloves, which is always detested. Since my hands are free, I mark the hourly count and record the environmental data. As soon as I get my sandwich and chips out and get comfortable, just as my mouth touches the food, a bird appears in the sky. The bird is far away, poorly lit, and is not obvious. Clearly, I will need the spotting scope to have a chance for identification.

I quickly put down my sandwich, jump to the scope, and work out the identification. This happens every time I attempt to eat. Even on birdless days, I can count on something happening that will necessitate action, just as I begin to eat. One day, it was a strange rustling of leaf litter in a crevasse behind and below me. It turned out to be a porcupine attempting to climb up the cliff wall. Another day, it was a U.F.O. The unidentified flying object turned out to be a plastic bag that was lofted high into the air. Today, an American Kestrel is the suspect. Because of their small size, they are sometimes difficult to detect and always difficult to photograph.

Photo by Brian M. Wargo

An American Kestrel fans its feathers.

During this lunch hour, a murder of crows passes the site. They are not migrating and will most likely return later in the day. A mixed group of Black Vultures and Turkey Vultures appear in the sky; they also are not migrating. They eventually land on the rocky outcroppings on the cliffs. This is one of their favorite spots, and the rocks are stained with their white excrement. Their rest is disrupted by a local pair of Common Ravens. I hear the ravens croak before I spot them. I croak back and get their attention. One, flying under the rock I am situated upon, deviates its path and comes back to investigate. It is disappointed when it sees me instead of another raven. It turns, rejoins its partner, and heads towards the vulture.

The vultures, knowing what is coming, take early flight to get a jump on the ravens. To no avail, the ravens chase the vultures, swooping down on them, making them roll and turn in the air. It reminds me of a dog running into a flock of pigeons in a city park. The difference is that the dog will then indignantly roll over and lick its anus, thereby revealing why it is just a dog. The ravens, when done demonstrating their dominance of the cliffs, will instead build some tool for future use, thereby showing why they are the kings of the mountain.

About an hour after lunch, I drink my hot tea. I am always amazed at how efficiently the thermos keeps the tea hot, and am equally amazed how wind can cool that tea when using the open mouth cap as a cup. The large surface area of the cup loses its fastest, hottest molecules from the surface, making hot tea into cool tea in just a few minutes. Like clockwork, birds seem to appear in the minutes when tea is served. I am excited about the prospect of drinking hot tea and seeing a migrating raptor, while simultaneously worrying about my tea getting cold and losing the bird. *"I should eat lunch and drink tea all day long,"* I jest, especially on slow days. But, it doesn't work that way—I've tried.

By afternoon, the temperatures usually rise, and if the sun can, it makes its presence known. The warm sunlight is a gift, heating my black coat and black pants. It is tempting to take off a layer, but I know better. The wind always wins this battle. It's best to conserve as much heat as possible. As the day gets warmer, I still get cooler. My body has been burning fuel to keep me warm all morning. Lunch and hot tea have replenished my reserves, but I am not moving around much, just standing. I, therefore, remain bundled, complete with gloves and a facemask.

This is the time of the day when visitors tend to arrive. Having just walked up the hill, they are warm, often sweating. They trepidatiously approach the site, curious to see why someone garbed in a snowsuit with optical equipment is standing on the edge of the cliff. I know they are thinking this because they usually ask me this question. Similar to lunch and teatime, when a guest arrives, a bird often appears. Not understanding birding etiquette, where bird identification trumps conversation, I patiently wait for a pause in the conversation before ignoring the visitor and perusing the bird. I desperately try not to be rude, but I usually only last about half a second before I just can't wait any longer and switch my attention to the bird.

I talk the visitor through the identification process, pointing out that the birds use the ridges as a flyway for their migration. The guests are usually intrigued by the thought of an interstate highway in the sky that raptors and other birds use to get from points north to points south. Their eyes widen and they exclaim their sudden awareness with a "WOW!"

Some visitors, of course, specifically come for the hawkwatching. They know about the site from viewing hawkcount.org, which lists all of the working hawk sites in the United States. More than likely, they have been following the daily tallies of this hawk site as well as others in the area.

Visitors rarely stay long, and before I know it, I am again alone on the cliff. The afternoon hours can be exciting, tiring, or boring. When counting hawks, there are no definites. When birds should be flying, they aren't. When the weather seems too dreary, they are. I long ago gave up any hope of predicting bird flights on any particular day.

Predicting the birds is like predicting a roulette wheel. I know that, given enough spins, particular numbers will be hit. For birds, I know during particular weeks, certain birds will inevitably migrate through. Unfortunately, for the hawkwatch that I am manning, a counter is not present every day. Therefore, many high number bird days are simply missed.

On this particular day, I feel like the Golden Eagles, which should be migrating at this point in the season, are in the air and will be arriving any minute now. While this idea seems reasonable, it has no correspondence with reality. However, Golden Eagles tend to be seen more in the afternoon at this particular site than in the morning. A more accurate statement is that Golden Eagles do not seem to be early birds.

As I ponder the many reasons that Golden Eagles should be passing the site, an adult male Northern Harrier makes his way south. He is spectacular, with his very long wings and tail, his ghostly owl-like pale face, gray body, and black wing tips—nothing looks like a Northern Harrier. He is far away, but I am still excited to see him flapping south. Within a few minutes, a few Red-tailed Hawks also pass south. Then a Sharp-shinned Hawk and a Cooper's Hawk appear. The Sharpie is migrating, but the Cooper's is a local. I regularly see this bird, which flies past the cliff, usually staying low and headed in the east-west direction.

I mark the log sheet and check the environmental measurements at the top of the hour. It is 3:00 p.m. Eastern Standard Time. In bird time, it is 4:00 p.m., for birds, like the residents of Arizona, ignore this vestige of agrarian life. Only humans are mad enough to think that they can control time. I watch the skies for another hour, until 4:00 p.m. Eastern Standard Time or 5:00 p.m. bird time. No birds appear in the sky during this time.

The sun is dimming as it lowers in the sky. I pack up and begin the walk down the hill; I am tired, wind-blown, and hungry. I contemplate if I should stay another hour but remind myself that no birds flew during the last hour, that I have children I would like to see, and that I am ready for dinner. I also remind myself that I will be back here tomorrow, and that the birds will surely wait for me to arrive. Or maybe, they will all stream through the moment that I leave. I stop … I think … I choose my family. The birds will have to wait until tomorrow.

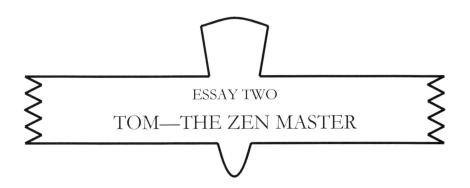

ESSAY TWO

TOM—THE ZEN MASTER

I know I am close to my destination as I pass a gothic-looking eagle statue. I can barely see it as I turn a corner of the one-lane road meandering through a parcel of dense deciduous forest. The statue is imposing and slightly eerie, for it is postured in a manner that has the eagle looking down on anything that passes this point. The morning fog is pervasive, and the sun has yet to make its presence known. The statue, the darkness, the fog, and the desolate road make me feel like I am in Transylvania.

I stop in an open space that must be the parking area. The wind is blowing from the east. I can tell by looking at my compass and watching the movement of the thick fog. I get out of the car. It is a surreal experience, like being at a rock concert standing in front of a thousand smoke machines. Unlike a concert, however, silence dominates this space. I feel slightly disoriented. I begin to walk slowly. I feel like a drunken sailor awakening after a bender. I am tempted to find my boat that must be somewhere out there in the fog. *Where am I?*

Reality returns. I am not a sailor. I am not in Boston or at a rock concert. I am in the central mountains of Pennsylvania. I have just arrived at the Allegheny Front Hawk Watch early in the morning on this fall day. Within a few minutes, clarity returns. Some of the fog begins to dissipate. I begin to walk east from the parking area. As I ventured out into the fog, I hear a voice yell out, "Past the barrier of thickets, the hill gets really steep, and things get much more dangerous. One member, years ago, broke his ankle down there." The voice was of Bob Stewart, who was the official counter of the day. Sunday is Bob's day, and his voice became a regular part of my weekends.

The Allegheny Front Hawk Watch is located along the eastern edge of the Allegheny Plateau, which is a rare flat spot in this otherwise mountainous terrain. Looking outwards from the edge, the next set of ridges are ten miles directly to the east, with a large valley between. Getting here requires driving up some rather steep roads. Depending on your starting point, the roads may be paved or dirt.

The dirt road is a more direct route to the hawkwatch site, but is precarious. Most drivers take the long way around the mountain to avoid this sometimes impassable road. If there is snow on the ground, this dirt road, called Shaffer Mountain Road, must be bypassed. The paved roads are safer, but also become treacherous with snow. On this day, the roads were moist with dew, and Shaffer Mountain Road was not only passable, but was also populated with a few Ruffed Grouse.

It was about 8:30 a.m. on Sunday when I arrived at the site and heard Bob's voice. He was already pulling out the paperwork for the day as he spoke to me. "The wind should pick up this afternoon," he gently chuckled. Bob has white hair, wire-rimmed glasses, and a slightly wily laugh. He is a pharmacist at the local drug store, and he also functions as the membership department for the Allegheny Plateau Audubon Society.

As I stood next him, I looked out to see the entire valley below filled with clouds. It is like standing on the edge of a great lake, with great billowing clouds filling the vast gap between you and the next tall ridge. By 9:00 a.m., the sun began to creep over the ridges and in a short time, a remarkable transformation occurred. The sun burned off the fog, and within an hour, the entire valley below was exposed.

It was about 10:00 a.m. when Tom Dick, the founder of the Allegheny Front Hawk Watch, arrived. Technically, Tom Dick is Dr. Dick, D.V.M. He holds a veterinarian license, but has spent most of his time in the last 20 years working as a naturalist. He purchased and then converted a run-down farm into wetlands. This was the natural state of the property; Tom just reverse-engineered the draining process. Nature did the rest.

Tom asked about the bird count so far, and Bob let him know that despite not seeing any raptors yet, it should be a pretty good day. "The winds are supposed to be out of the south in about an hour according to Weather Underground," replied Tom. South winds are good news at this site. East winds are great news. West winds are the norm, and that, unfortunately, pushes the birds out away from the hillside, making it difficult to find and count them. Tom stated that he had work to do at the wetlands but would be back in the afternoon.

Tom takes great pride in the wetlands that he rejuvenated. His vision had been a remarkable success. By transforming an unproductive farm into wetlands, the number of species on the 241-plus acres skyrocketed. Many species are now breeding there, including Hooded Mergansers, Blue-winged Teal, and Wood Ducks. In addition, over 25 species of waterfowl regularly stop at the wetlands during their migration. For his vision, commitment, and determination, Tom has been honored with multiple awards for his conservation work. He is also known throughout the region for his stance against the wind turbine industry, who proposed putting turbines in the migratory path of the Golden Eagle. To some locals, Tom is a hero. To others, he is just a really smart guy who knows a lot about animals.

Tom owns the land of the hawkwatch site, which he bought over 20 years ago. He has been improving it ever since. This includes not only the flora and fauna, but also the human resources. Over the years, the Allegheny Front Hawk Watch has built up a steady and dedicated following. There are over 10 potential counters that can count on any given day. Each season, the counters volunteer to take a particular day of the week. Having extra counters always leaves a substitute who can cover for another counter if someone has to call off. It is like a small army of volunteers, and Tom runs the show.

Tom's dedication and hard work should be commended, but what Tom has really done is create a place of Zen. Tom lives out here in rural Bedford, where he is surrounded by the serenity of nature. Zen is not just a metaphor, for Tom is a practicing Buddhist. When not finding nirvana, Tom treats animals, lectures at the local university, and works as a part-time farmer. He looks good for his age, which could be attributed to his daily regimen of running five miles.

Bob and Tom are just two of the regular members who count hawks at the Allegheny Front site. Others include Rosemary McGlynn, Ed Gowarty, Jim Rocco, and Bob Gorsuch. Ed and Jim seem to show up most days and are steady, reliable hawkwatchers. Both are quick with stories, but Ed sometimes brings homemade wine for the group. Jim ribs him about the taste, but it is in good spirits. They are both good guys to be around. I always stand next to one them, because even when talking, they always have an eye to the sky. Ed and Jim each take a weekday to count, but it seems as if they are at the hawkwatch site regardless of the day. This helps when a surge of birds comes through or new arrivals need help understanding the purpose of the hawk count. But, if you want a greeter, Jack Julian is your man.

Jack is the treasurer of the Allegheny Plateau Audubon Society and is a regular at the hawk site. He is a full-time Physics teacher at the local high school, so Jack is not (usually) at the hawk site during school hours. He is a Jack-Of-All-Trades (pun intended), a true generalist. He knows a little about nearly everything. On any given day at the hawkwatch, there is a chance that Jack will show up, usually with a particular task that needs attention. It may be putting a battery in the remote environmental transmitter, gathering paperwork from a previous counter, hauling supplies for an upcoming picnic, or just helping to spot birds. Jack is friendly, active, and always there to greet visitors to the site.

My favorite attribute of Jack is his raven call. My daughter, Meadow, also has a good raven call, but Jack's is more baritone. It should be noted that Jack is a large man. He knows how to project his voice, being a teacher and member of the local choir, so my daughter really doesn't have a chance against Jack. Despite his size, Jack is one of the most active people I know. He is always biking, kayaking, birding, singing, visiting, helping, camping, volunteering, or teaching.

One other note about Jack: When spotting birds, Jack is one who always lets you know which bird he would like it to be before actually identifying it. This is usually a faux pas for any hawkwatcher, but there is never a foul for Jack. If you ever visit Allegheny Front Hawk Watch, make Jack Julian your first acquaintance. He can speak intelligently to anyone about anything and usually knows something about any topic you mention.

On the opposite side of the spectrum is Bob Gorsuch, the Monday counter. He probably sees the least visitors and certainly the smallest crowds. Gorsuch, an avid hunter and former trapper, seems to have remarkable eyesight for his age. He is one of the few at the site who does not wear glasses. "Mondays are always slow days unless there is an east wind; then, there are always visitors," Gorsuch tells me on a particularly slow Monday. It was just one of these lonely Mondays that Gorsuch was sitting at his usual spot in the field when two black bears walked right up to him. "From that day on, I always carried a gun," he reminisces.

This is not the first bear incident at the hawkwatch. Ed Gowarty was scanning the sky using his binoculars when he heard something pass him on each side. Unfazed, Ed just kept looking intently through his binoculars. When he finally put down his binoculars, he looked behind him. There, running into the woods, were two black bears. "I thought it was two deer who got scared, and thought nothing of it. But when I saw it was bear, my heart sank a little bit."

Tom Dick has also had a run-in at his house with a goliath black bear, who was after birdseed. "It was so dark when I went out to investigate the noise that I almost walked into the bear," recalled Tom. Just about every one of the counters has a bear story. Thankfully, all ended without incident.

But our lives changed at the Allegheny Front when we met Deb Bodenschatz, a salamander expert turned hawk counter. Deb's day to count was on Saturday, and we soon found out that Deb is a remarkable host. Each Saturday, she would serve homemade hot food in the middle of the field—delicious marinated steak sandwiches, pastries, homemade trail mix, etc. It is just incredible! Her ability to organize and manage is spectacular. She makes it seem effortless. My son, Theo, is allergic to peanuts, so Deb converted to all peanut-free dishes, which she said was "no big deal." I can tell you that, for the average cook, going peanut-free *is* a big deal. Not for Deb. She is the Martha Stewart of hawkwatchers. She is amazing!

No matter who is counting for the day, there are bound to be at least a few good birds. Some days are slow, as is the case when the winds are from the west. However, when birds are present, the views are spectacular. Other hawkwatches will have higher numbers of birds on most days, but it is hard to match the quality of the views at Allegheny Front.

Some days are more than spectacular—they are legendary. Like the day that 74 Golden Eagles passed in a nine-hour period, or the day when nearly 8,000 Broad-winged Hawks flew past in just a few hours. In addition to the raptors, other birds are usually flying, either from the wetlands or from Shawnee Lake, both of which are visible from the hawkwatching site. A steady stream of non-hominid and non-avian visitors helps keep the day rolling, especially on slow days. An occasional Garter Snake, Praying Mantis, Grey Squirrel, Pileated Woodpecker, Walking Stick, Monarch Butterfly, Garden Spider, or an Inch Worm always seems to lighten the mood.

Exotic hominid visitors also come in a steady stream on the weekends, usually during the migration peaks. Mushroom experts, butterfly specialists, wine makers, salamander scientists, wildlife managers, graduate students, power plant owners, Mennonites, and specialty photographers may show up on any given day. The conversations often leave the other members wiser than when they came.

Photo by Brian M. Wargo

A Pileated Woodpecker flies by the hawk site at Allegheny Front.

One of my favorite specialists is Dave "The Owl-Man" Darney. Dave is a full-time electrician who happens to be one of the few owl banders in the state. Watching this big, strong, rugged working man daintily handle the smallest owls east of the Mississippi is a sight to behold. He is so delicate with the owls that it is makes you wonder why they dig their killing talons into his hands. Dave never flinches; he just ignores the pain. This may be an unwritten passage to becoming an electrician—having strong, pain-resistant hands that, as Dave states, "Can put them across 110 [volts], no problem."

Owl banding complements the hawkwatching at Allegheny Front. As the day of watching hawks winds down, work begins in setting up nets for owl banding. Depending on the number of volunteers available, the nets can be hung within an hour or two. The owls are not active until dusk, so there is a lull in birding in the evening. But, then it happens—a transformation to something akin to an '80s movie about Vietnam. A loud speaker begins to blare the monotonously repeating call of the Saw-whet Owl: *Too-too-too.......* *..... too-too-too............. too-too-too............ too-too-too*, repeated over and over and over.

Photo by Brian M. Wargo

Dave "The Owl Man" Darney inspiring another
generation of conservationists.

If you want to psychologically destabilize someone, this is a great method: place them in a dark forested area that is nearly pitch black, cold, and disorienting, and then play the Saw-whet Owl call at full blast every five seconds. The reason that people willingly subject themselves to such torture is that, every 20 minutes, the nets are checked. When conditions are just right, owls miraculously appear, tangled in the nets. Ordinarily, the chances of seeing an owl are miniscule. Here, not only can you see an owl, but you can examine it up close. It is a truly remarkable encounter. But, do not get too excited, for Dave is a serious man that will not allow any activity that stresses the owls. Listen to his instruction, remain quiet, and you will be rewarded with a once-in-a-lifetime experience.

Once caught, each owl must be weighed, measured, sexed, and given a quick physical before being released back into the night. Time is of the essence, for owls that are not moving begin to cool rapidly. Therefore, the owls must be processed and release within minutes of capture. All of the owl data, as well as the hawk data, is in service of helping wildlife scientists better understand these marvelous flying creatures. What is most striking is that the general public is allowed to participate in this citizen science activity.

The Allegheny Front Hawk Watch is a unique hawkwatch. You can tell how special this place is to so many people by attending the summer or fall picnics. As many as 50 people might show for each event. The day centers around hawkwatching, but many of the older hawkwatchers and way-out-of-towners tend to show up for these gatherings. Sometimes, members from over 25 years ago, attend. With them, a set of historical tales about the early days of the hawkwatch, thus providing perspective to its origins.

It is a triumph of hawkwatching to get so many people together to celebrate the birds. It could almost start a new holiday: Hawkwatch Day! If you think that this is hyperbolic or you are new to hawkwatching, I highly recommend visiting the Allegheny Front Hawk Watch. You will always have a spectacular view, encounter kind people, and, hopefully, experience some great birding. You may even find yourself in favor of creating Hawkwatch Day.

ESSAY THREE

BECOMING A COUNTER

The genesis of my hawkwatching bug began with general birdwatching. Being outside and becoming aware of the natural world was sedative and addictive. More of my time and energy flowed, unconsciously, in this particular direction. At first, I thought it was about seeing new places, but I soon realized it was about the birds. Holiday and summer trips were planned around birding areas. Soon, every weekend included birding activities. Before I was aware of my slide into it, I became a birder.

Hawkwatching was a natural extension of general birding. A little research online revealed that birders were attending hawk migration sites during the fall season. They were located along ridges on the east coast. The closest set of ridges to my home in Pittsburgh was the Allegheny Front. I would later learn that this hawkwatch is sometimes called the "Cadillac" of hawkwatching because you can drive up to the site, walk about 50 ft. from your vehicle and observe birds that are swept up to you, yielding incredible views.

In addition to the birds, a group of regulars attended the site. They were the first hawkwatchers I encountered. Many of them were counters and some showed up just about every day. It became apparent that for many, this and other hawkwatches served as the centers of the hawkwatching community. Some hawkwatchers were loyal to a particular hawk site while others frequented several sites. I think it was the camaraderie coupled with the landscape, and of course, the close-up views of the birds that lured me in and made me a hawkwatcher.

The draw of counting hawks is most likely tied to my training in science. Working on my Master's in Physics transformed me into a hypothesis generating, data collecting fool. Knowing if patterns existed in nature was always tempered by the need to gather data. When opportunities arose where data could be collected easily, I would collect it and analyze it, usually just for pleasure. My training then turned to working on messier data sets, that of human cognition. For ten years, I worked on a Ph.D. in Instruction and Learning – Science Education. Most of that time involved research in the most chaotic of systems, human beings and how they learn science.

Counting birds that happened to fly along a particular cliff during migration season was as therapeutic as it was exhilarating. When I needed a break from academic work, I would bird. More often, counting birds meant counting hawks. Unfortunately, guilt was always birding next to me, proselytizing, *"If you have time to watch birds, you clearly have time to write, analyze, or inform others of your academic work."* I would temper this guilt with the notion that I was helping the raptor community by occasionally yelling, "Bird!" thereby causing the official counter (and other able hawkwatchers) to zoom in on the bird, identify the bird, and subsequently count the bird. This record would eventually be shared with the rest of the hawkwatching community (and the public) on hawkcount.org. This justified my leisure time watching birds. I was helping with citizen science. Of course, deep down, I realized that I was not really pulling my weight. The counters would probably see the bird on their own, without my help. More importantly, the counters could *reliably* identify each bird.

I was earnest, committed, and determined to help identify the hawks. The problem, of course, is that the birds are often far away, moving fast, and gliding, thereby hiding their characteristic identifiers. To help become a competent hawkwatcher, I read and studied in detail all of the books on hawk identification, including *Hawks in Flight* (Dunne, Sibley, and Sutton), *Hawks at a Distance* (Liguori), and *Hawks from Every Angle* (Liguori). Each picture and diagram was annotated; my notes filled the margins of each paragraph. Supplemental books were used to get a different view of raptors, including *Peterson's Field Guide: Hawks* by Clark & Wheeler, the *National Geographic Field Guide to the Birds of Eastern North America* by Dunn and Alderfer, and *The Stokes Field Guide to the Birds of North America* by Don and Lillian Stokes.

In addition, I downloaded guides and PowerPoints from various hawkwatches (*Learning Eastern Hawk Flight Identification* by Bob Pettit at Holiday Beach is particularly good). I quizzed myself by blocking out the descriptions of the birds in my books, testing to see if I could arrive at identification. (This was before the *Crossley ID Guide on Raptors* by Crossley, Liguori, and Sullivan, which seemed like it was written just for me.) I would go online and try to identify raptors, but became frustrated with the misidentifications found there.

After quizzing myself, studying the guidebooks, and preparing intently, I would confidently arrive at the hawk site, only to freeze with each passing bird. All of the studying and practicing was insufficient for the very quick views of the birds. By the time I had locked my binoculars to the bird, focused, and picked out a descriptor or two, the bird was gone, and with it, its identification. It seemed hopeless. *"Hawk counters are simply born with an innate ability to identify hawks!"* I flippantly noted.

Eventually, I was able to identify the raptors. What helped the most was standing near seasoned hawkwatchers and speaking to them as birds passed. Most would answer my questions about the hawks and would even allow me to try to speak aloud as I thought through the identification process. They would then point out the fine details needed to identify the birds. In effect, I was informally apprenticing under the guise of the more experienced hawkwatchers. Waggoner's Gap, Ripley, and of course my home hawkwatch, the Allegheny Front, were gracious hosts for a developing hawk counter.

Photo by Brian M. Wargo

Beginners find it difficult to identify hawks, even when they are close, well illuminated, and staring back at them.

My experiences observing and unofficially apprenticing near master hawkwatchers has led me to believe that master hawkwatchers identify birds differently than how novice are taught to identify raptors. Master hawkwatchers seem to identify raptors based on essentialism, while novices concentrate on characteristics. Below is my analysis of this process.

Expert hawkwatchers seem to create an idealized bird in their mind, one that is animated, that shows vitality, and is discretely different than all other birds. Instead of building a bird from the characteristics displayed, they match the bird to their idealized mental bird and then check to see the level of fit. This gestalt quickly aligns their conceived bird with the actual bird.

This process works not from deconstruction of the bird, but holistically, assessing the totality of the bird's characteristics all at once. This process is automatic for master hawkwatchers and is why novices find their abilities mystical. This level of expertise allows time to focus on the minutia of feather length, color, and molt, as well as flight characteristics, behavior, and disposition. It is these details that are often vocalized, thereby giving the *impression* that master hawkwatchers move linearly through a flow chart to accomplish identification. In reality, their identification happens unconsciously and is nearly instantaneous. They then use the fine details to strengthen their confidence in their identification.

I want to stress that this is normative behavior for expert hawkwatchers and is not prescriptive for becoming an expert hawkwatcher. There may be no shortcut for learning to identifying as an expert. It may be necessary to at first concentrate individually on plumage, shape, and proportions. As each is mastered, they may become less cognitively burdensome, thereby allowing the mind to concentrate on other aspects of the migrating raptor.

In time, the initial ways of identification fade as more sophisticated techniques are employed. This is similar to the rigging and scaffolding required when building a bridge over a large span. At first, the bridge is held in place by cables and a chaotic array of supports. As the bridge becomes structurally sound, the scaffolding is taken down, leaving the freestanding structure. It would be nice to have a fully formed bridge and simply place it where it is needed, but our construction equipment is just not capable of such a job. Similarly, it would be nice to identify raptors as an expert, but our cognitive equipment forces us to do so piecemeal, at least at first. As we gain experience, our initial ways of identification are forgotten as we develop a more streamlined approach.

This process takes time—a lot of time. Studies of master chess players have shown that it takes tens of thousands of hours to achieve masterful play. Even then, not all who play that many hours will necessarily become masterful. Therefore, it is not just the time at play that matters, but also, about how they play. *How one practices can impede or progress development.* This analysis may explain why some hawkwatchers, who have spent copious amounts of time watching raptors, remain unreliable with their identification. This also explains why attending identification workshops can improve our identification abilities—for it often teaches us *how* to practice.

Photo by Jeanine Ging

Expert hawkwatchers have an essential image of the hawks in their mind.

In the end, it is not possible for anyone to identify every raptor, every time. As Kenn Kaufman (author of the Kaufman Field Guide Series) stated during one his talks, a birder can become paralyzed if they are afraid of ever making a misidentification. Heeding Kenn's advice, I always try to identify every raptor, every time, but then I ask about my confidence in my identification. If I feel that nine out of ten other hawkwatchers would independently arrive at the same identification, I say that the bird has been identified. Some hawkwatches at certain times have this luxury of having well over ten masterful hawkwatchers viewing the same birds. More likely, for most hawkwatches, a group of five total hawkwatchers of varying ability is a good day.

For the past few years, I devoted my hawkwatching hours to a newly formed hawkwatch in Cumberland, Maryland. This site helps fill in a gap between other hawkwatch sites that are on ridges to the east and the west. The goal is to provide complementary data to develop a better understanding of how the raptors use these particular ridges as they migrate. To collect this data requires an arduous hike up a long steep hill to the site, which is precariously situated on the edge of a natural cliff. The fitness required to get to the site, as well as the lack of facilities once at the site, are impediments to attracting other hawkwatchers and counters. If the journey is made, however, the panoramic view alone is sufficient reward. And of course, the birds, for the natural cliffs attract many hawks, including Peregrine Falcons, which spend time playing, hunting, roosting, and harassing other birds.

As this fall migration season comes to an end, and I reflect on my first solo year as a counter, I recognize that being able to identify hawks is only one aspect of being an official counter. The commitment required can be taxing, especially for non-established hawkwatches. Attempting to work out the logistics of transportation, land rights, data entry, public relations, and scheduling can be daunting. Established hawkwatches, over the years, develop regulars who attend for the viewing of hawks, but also in mediating the multitude of small projects and remedying the small problems that arise.

The overarching concern of any hawkwatch is in manning the site. This is the most difficult duty to fulfill, for most hawk counters are volunteers. They have lives, other commitments, and families that limit their ability to dedicate an entire day to counting hawks. Even if they can dedicate one day here or there, most likely the hawkwatch will remain barren some days unless a large coalition can be formed. Even if seven individuals dedicate time to a particular site, that requires one day a week of bird counting from each individual from late August through December. And that is just for the fall count.

The term "volunteer" does not diminish the level of pride, dedication, and effort each birder devotes to getting it right each time they identify a bird. The level of concentration and attention required is impressive. Some days, a flurry of birds will keep a counter busy most of the day. Other days, a counter may sit for hours, sometimes alone, without any birds in the sky. This stillness can be punctuated by a quick burst of birds flying overhead, requiring the hawkwatcher to spring back into identification mode.

In addition to the discipline of watching for birds, often for seven, eight, or nine hours at a time, there is a physical component to hawkwatching. Most hawkwatches are precariously placed on rocky cliffs, hills, or mountains, requiring hawkwatchers to carry their equipment and supplies each trip. This physicality is often exacerbated by the age of the hawkwatcher, who tends to be older than a spry twenty-something (or even a thirty-something). Older hawkwatchers are often retirees who are able to fill in during the week, when younger hawkwatchers are still building their nest egg.

The environment that hawkwatchers work in is one of difficulty, where wind, sun, and cold temperatures collaborate to make all aspects of hawkwatching arduous. Thinking quickly and lucidly when the body is cold is difficult, but pales in comparison to writing legibly with cold hands. Yet, it is often senior hawkwatchers that are counting during inclement weather. So, is the stereotype of older aged individuals being more sensitive to the elements just a myth? In reality, senior hawkwatchers are probably more sensitive to the cold, owing to the fact that as the body ages, metabolism slows. The all too common sight of a student at a bus stop wearing shorts in winter, or of a grandmother wearing a coat in the middle of summer are illustrative examples of this aging process, and rarely requires narration when seen. Yet senior hawkwatchers often man the hawkwatch sites during the cold weather of November, December, February, and March.

The description of hawkwatching above paints a picture of an activity that is laborious, challenging, often uncomfortable, and usually devoid of recognition. So why would anyone count hawks? I like to tell myself that it is because I am contributing to the wealth of scientific data on the birds. And that is true. Without citizen scientists making observations and collecting data, scientists would not have sufficient information to make global connections about the current state of the hawks. But, to be honest, I think most hawkwatchers view the birds mainly for the pleasure of seeing magnificent creatures doing what we are not able to do.

The act of hawkwatching seems to satisfy the portion of the brain that seeks reformation. Religious individuals seek a weekly gathering to rejuvenate their souls. In a similar manner, hawkwatchers connect with their inner biophilia, a term coined by E. O. Wilson that expresses the innate attraction and wonderment of the natural world and especially, its living systems.

In my first year as a counter, I logged over 130 hours of counting hawks. When questioned about why I do it, I often respond using scientific reasons. The real answer is more complicated and would take too long to explain. Given sufficient time, I could clearly lay out a logical case for watching hawks—at least that is what I keep telling myself. What I can say is that counting hawks brings satisfaction. Overcoming the physical and mental barriers, both of which are daunting, gives a sense of achievement. Learning a craft, practicing it intently, and pulling it off is anything but accidental. It is what disciplined people do! Why we do it...that remains to be explained.

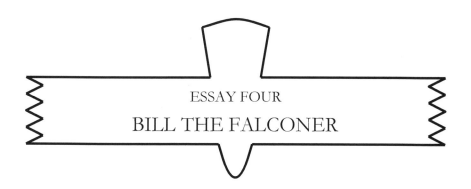

ESSAY FOUR

BILL THE FALCONER

Being at the right place at the right time is an apt description of my day with the falconer. I was looking into a new roadway that leads up to a hawkwatch when I ran into a guy who knew a guy. Within an hour, I was walking with a falconer who was setting up a spring-loaded hawk net. The goal was to trap a Red-tailed Hawk for a beginning falconer who was apprenticing under the master falconer, named Bill. It was Red-tail migration season, and the day looked promising.

Bill Tierney was the first falconer I had ever met. I had only heard stories about falconry, but had no direct experience with it. My naïve understanding of the practice involved a shaman who somehow developed a special bond with the hawks. It seemed mystical that a hawk could be coaxed into hunting for and then returning to a man who would keep it captive.

"Why would a wild, free-flying bird willingly give up its freedom to work for a man?" I wondered. Clearly, that man had to be special!

Bill and the apprentice did not seem special, nor did the third falconer who was helping for the day. These men seemed like normal people, like hunters, but more sophisticated. All three men were a little apprehensive when I first arrived at their site. They heard that I counted hawks under their guise of Hawk Migration Association of North America (HMANA) and that I was a member. The man who had introduced us owned some of the adjacent land next to the field in which we were standing and figured it would be a good idea to put us together.

When I introduced myself, I was up front about my goals for counting hawks, stating that I was part of a citizen scientist group that helped to provide data about migratory raptors. I let the three men know that I was interested in understanding falconry and had no ill intentions about this meeting. I filled them in on my science background and my desire to learn more about how they birded. All three men cautiously explained their goals for the day and some basic facts concerning falconry.

After a short discussion, I could tell these falconers were anxious to get started. They had driven two hours to this site and, clearly, they would rather be setting up than schooling me on falconry. Finally, I asked if they would mind if I spent the day with them, expecting that they would find a polite way to reject the idea. Surprisingly, they said that would be fine. I was excited and did my best to be helpful. I carried equipment, helped set up tents, and tried to stay out of the way when not doing something useful.

When the set-up was completed, Bill welcomed me into his camouflage tent that functioned as a blind. Over the next six or so hours, I learned a great deal about falconry. I was intrigued and my questions were unending, but all three men graciously answered all of them. I could tell they genuinely loved this sport.

At first, they put the best possible spin on the practice, which I would expect when introduced to any new activity. That type of grace period only lasts for the first hour or so and then the men let down their guard a bit. They began to reveal a more realistic view of the sport, including the perception of it with the public. They were aware of the animal rights influence on popular opinion and the current movement towards a hands-off approach with all animals. The men also expressed the perceived attack on hunters from various well-intentioned groups.

Bill articulated to the apprentice the importance of always showing the sport in a positive light. He then told me that falconers seem to self-regulate, putting pressure on falconers who do not follow the rules and ethics of falconry.

"I personally will not hunt with certain falconers. It is like any other organization; you always have a few that give it a bad name."

Bill explained that falconry is heavily regulated and that falconers are subject to random inspections by the state. It was clear that Bill was a hard-liner on rules and enforcement. His disposition was that of a military man, rather than what you would expect from a registered nurse, which was his current occupation. Bill seemed to have a blend of genuine care coupled with a no-nonsense view of the natural world, society, and falconry.

The men spoke of their love of hawks without using the word love. They had a deep respect for the hawks they captured and kept. At the same time, they tried not to become attached to the birds nor anthropomorphize the individual hawks. Bill explained that most of the hawks will spend the winter and spring with the falconer, but will eventually be released back into the wild. I asked if that was hard after working for so many months together. All three men stated that they try not to get attached to the birds, but it happens. The third falconer told of a hawk that he released, only to receive a phone call a week later that his bird had been hit by a car. He left a tag on the bird's leg that identified him as the bird's falconer. He said that was really tough. He then reminded me that most hawks do not survive in the wild, with 70% not seeing their second year of life. That rate changes to 90% by year five.

Bill made it clear that hawks are not pets. They do not have a natural propensity for the social. The hawk is a temporary member of a symbiotic hunting team. The hawk only cares about food and could care less about people.

"People like hawks more than hawks like people," I said. Bill nodded in agreement.

Bill forthrightly describes falconry as a blood sport, one in which animals die.

"Lots of people love to see these birds, but they don't really think about how they eat. These birds catch and tear apart other living things." Bill's comment is an important one and brings up an interesting conundrum of hawkwatchers, who are often birdwatchers. The hawks that birdwatchers love to watch kill the other birds that birdwatchers love to see. Bill and I agree that people are often disconnected to the part of the food web that includes carnivorous species. Humans can be thought of as such a carnivore, but we are rarely involved in the act of acquiring the flesh that we consume. Hunting and falconry directly links humans back to the killing of the food they eat.

Falconers are intriguing, but not simply because they are aware of where meat comes from. They must adopt a particular lifestyle.

"This is not an activity for everyone. Most people do not understand the commitment and the dedication required," muttered the apprentice as he set the trap. He then described the hours needed each week to maintain a safe and healthy habitat for the hawks, as well as the hours needed for training and maintaining a working relationship between the hawk and the man.

"The economics of falconry is also a real factor," chimed the other falconer. He continued, stating that building a housing unit, budgeting for food, buying gasoline for transportation of the birds, and traveling to meetings all require money. I then stated my conception of falconry:

"In many ways, falconry is analogous to parenting; the welfare of the bird is the responsibility of the falconer." All three men quickly affirmed the truth of my statement.

I get their point, for just watching hawks can be an expensive, time-consuming, and arduous process. In the end, I am passively describing the birds that I see as a hawk counter. I enter numbers onto the hawkcount.org site, and then I am done with my duties. I am only responsible for myself. In contrast, the falconers must worry about the safety and condition of their birds. This paradoxically includes the bait birds, which are usually pigeons.

Most falconers keep pigeons. They can be used to attract hawks to the trap. The pigeon is placed in a harness and is tied with a small string near the trap. This entices the hawk to come in towards the trap. My first time seeing the pigeon used for bait was a bit surreal. Bill loaded one of his "favorite" pigeons into the hawk trap, speaking to the bird as if it were part of the team. This produced ambivalent feelings, for the pigeon in the trap seemed like a nice bird, sitting calmly in Bill's hand as he placed the harness around it. I must admit, seeing dozens of pigeons under a city bridge usually invokes a visceral and repugnant response from me, but, here, the bird seemed rather nice. It was about to be sacrificed so that a hawk could be gained.

I asked Bill about this, and he said, "That pigeon? Oh, he will be fine."

"Fine?" I thought, *"How will he be fine?"*

After the trap was set, I got into the tent with Bill. The other two men were in a tent right next to us. Each tent was in charge of a trap. There were two small ropes that extended from the trap to the tent. One was connected to the bait pigeon. The other rope was the trigger of the trap. Bill could pull on the string that was connected to the pigeon, thereby keeping the pigeon active. Bill allowed me to control the trigger rope. My instructions were to pull the rope when he gave the word.

Within a few minutes, a Red-tailed Hawk came into view. It looked like it was migrating, but it suddenly stopped its motion across the sky.

"He's looking!" Bill yelled out. After about a minute, it happened. The Red-tailed Hawk decided to go for the bait. "He's coming in!" Bill yelled. The hawk that had been so far out just a minute ago was now approaching like a bullet. I could not believe how quickly he was at the net. "Get ready!" Bill exclaimed. At just that moment, Bill let out the slack on the pigeon line, who walked into a small safety box within the net. The Red-tailed Hawk landed and looked for the pigeon. Bill said, "Now!" I pulled the line and the net slowly and quietly closed upwards around the hawk. We then jumped out of the tent and quickly rushed towards the captured bird, slowing as we got closer so not to scare the perplexed hawk.

Photo by Brian M. Wargo

A very high Red-tailed Hawk stops in mid-air, skeptically viewing the falconers as they place a pigeon into the trap.

Bill assessed the hawk, giving estimates of age, sex, and overall appearance. He then put on a glove and spoke to the apprentice about the proper technique for safely securing the hawk. Bill emphasized the danger talons possess and used his other ungloved hand to distract the hawk. He slipped his gloved hand into the netting and the Red-tailed Hawk flipped onto its back. It held its talons out from its body, letting everyone know he meant business. With a quick and calculated motion, Bill grabbed the hawk right above the talons and held it upright. The apprentice opened the trap and removed the netting.

As Bill held the hawk, he checked the general condition of the bird. This particular hawk was heavy and in good condition. Bill extended one of the wings and looked closely at the molting feathers. He also pulled a few freeloading flies off the bird. The hawk was weighed and it was determined that this bird did not meet the criteria for keeping. This hawk was to be released, and Bill allowed me to do it. He instructed me on how to hold the legs, support the body, and keep the beak from getting too close to my face. With a gentle upward launch, the hawk took to the air and continued on its way.

Photo by Bill Tierney

Releasing a Red-tailed Hawk—a truly amazing experience!

During the few minutes that the hawk was in the possession of the falconers, the bird remained reasonably calm. I detected very little stress in the bird, other than occasionally leaving its beak open in a posturing manner. My non-professional, naïve, and unscientific opinion was that the experience had left the bird no worse for wear.

Clearly, the hawk would have preferred to not encounter four men and a net, but, overall, the capture and release seemed fairly innocuous. My condition, on the other hand, was one of pure excitement and joy. I had just encountered a raptor that was truly wild. I held this magnificent creature it in my hands, looked into its eyes, and had the pleasure of letting it go back to into the wild.

I must say that this experience was the pinnacle of my interaction with nature. While I do feel guilty in menacing with nature in any way, I hold this experience as one that has transformative potential. It is difficult to determine if I will ever look at a common Red-tailed Hawk in the same manner again. I have an even greater respect for the bird than I did before, and I am eternally grateful for such an awe-inspiring encounter. My gratitude for nature and all her inhabitants through this forced interaction will forever indebt me to her. I look forward to repaying this debt by being the best steward of nature that I can be.

After releasing the first hawk of the day, the trap was reset. The unharmed pigeon was pulled out of the safety box and commanded to resume its duties. I kept reminding myself that the pigeon was safe during the experience, but I still empathized with it. I could only imagine the feeling of being dangled as a food source for another animal. Surprisingly, the pigeon remained reasonably calm during the entire process and seemed no different after its first encounter.

After some time, several birds were migrating past. I was impressed by the identification skills of Bill and the other men. The birds were rather far away, and the sky was mostly clear. Here, I was able to help the men in identifying the birds in the sky. Their knowledge of the actual bird was superior to mine, which was mainly limited to description. These men have held many of these birds in their hands and may have even hunted with them. The stories they shared were mesmerizing. It was like knowing a trainer of a famous sports person, except we were talking about different species of raptors. Each type of bird has its own quirks, disposition, and issues.

Even the various parasites, ailments, and weaknesses of each bird were interesting to hear. The descriptions of how the bird and the man interacted were certainly my favorites. I have been watching hawks for some years and always wondered what they were like, how they would respond to an encounter with a human, or what they might be thinking.

The take-away from the conversations with the three men was that hawks, in general, have a one-track mind. Their objective is to find and secure food. That is their modus operandi. Nothing else mattered, or if did, it was minor compared to the primary objective.

I surmised by stating, "It sounds as if they are almost robotic killers!" The men laughed not only because of the statement sounded eerily sci-fi, but it was an apt description.

"Of course they're living creatures and they respond to things like living creatures do, but they don't sit around and think all day," one of the men responded.

"So they are the opposite of a raven," I said.

"Yah," they all chuckled in amusement.

"When you walk into their pen, they are not bored sitting there, cause they would be doing the same thing out of the pen. They would be waiting for their next meal. They are not excited to see you," the third falconer stated before I interrupted and tried to finish his statement.

"Unless you are bringing them their next meal."

"That's right," the third falconer said, nodding his head.

I asked, "If you had been working with a hawk for months and suddenly dropped dead, would the hawk be distressed or just begin to eat you?"

"It wouldn't be distressed," replied the third falconer.

"So, it would eat you?" I asked.

The third falconer chuckled, "Yah, I guess." Clearly, the four of us were oversimplifying matters, but I got the idea.

We went back into the tents, and I was starving from my exciting morning.

"What do you guys do for lunch?" I asked.

"We just snack and wait for the ride home," replied Bill.

"Do you mind if I eat?" I asked.

"Go for it!" replied Bill. I handed over my rope and pulled out a sandwich. Bill did not need me, of course, for he was the master; I was an observer. Immediately, a couple of Red-tailed Hawks appeared and two zoomed down towards the traps. The first Red-tailed aimed for Bill's trap. "Here he comes!" Bill yelped.

The pigeon did not move into the safety box. Bill yelled at the pigeon, "Get in the box! Get in the damn box!" The pigeon did not and the Red-tailed Hawk landed on the pigeon. Bill pulled the rope and the hawk was captured, but he had the pigeon in its talon. "Do you need a minute?" Bill yelled to the other men.

"Yah, the other one is still coming in," the other falconer replied. The other Red-tailed Hawk did come in and, just as it was about to land in the trap, it turned away and landed in a tree about fifty yards away.

"Let's see what he is going to do!" yelled the third falconer. After a minute or so, the Red-tailed Hawk flew away.

Bill said, "Let's get the bird," and we went out to the trap.

During all of this drama, the pigeon had been in the clutches of the Red-tailed Hawk, who was clearly confused and agitated by the netting around it. I feared the worse for the pigeon. As we approached the trap, I could see that the pigeon was still alive, but was being dragged around by the Red-tailed Hawk, who refused to let it go, despite its predicament. Bill was excited about the hawk, but was also worried about his pigeon. I wondered if Bill cared about the pigeon for its own sake or that this particular bird was just a really good hawk lure. I suspect the latter.

Bill put on both of his thick leather gloves and went in to grab the hawk and save his pigeon. The hawk was not letting go of the pigeon, so Bill had to open the hawk's talons. Surprisingly, the pigeon was in pretty good shape, just missing a clump of feathers on its back. This particular Red-tailed Hawk met the criteria and the three men decided to hold onto this one.

The men worked together to immobilize the hawk by holding it around its body. They put on a very small headpiece that covered the bird's eyes. The first hood did not fit perfectly, so they tried a second one. It was a nearly perfect fit. Bill very gently tied the hood in place. A stocking was then unraveled over the head of the hawk all the way down to it talons. The legs were secured using a small bit of tape. In the end, the bird was wearing what amounted to a see-through set of yoga pants for its entire body. Bill checked to make sure the fit was not inhibiting the hawk's breathing and that the bird was not exhibiting stress. The bird was then placed on small table and covered with a light blanket.

Photo by Brian M. Wargo

The perfectly fitting handmade hood instantly calms the captured hawk.

This process of securing the Red-tailed Hawk was peculiar. Here, the natural-born killer, the hawk, was being subdued by three men whose combined weight was a hundred fold of the hawk. Yet, the overpowering men were as delicate and dexterous with the bird as a mother nursing her young. The care and concern exhibited by the falconers for the hawk seemed odd after this bird just tried to kill one of the men's other birds. This confluence of violence and caring seems to symbolize man's struggle in this world, where we are not either primitive or actualized, but are simultaneously both.

The subdued hawk, after initially being pulled from the net and having its food taken from its talons, quickly vanquished its initial hostility. As soon as the hood was placed onto its head, the bird immediately was calm. After the stocking was placed around its body, the Red-tailed Hawk seemed to switch to standby or sleep mode. There was not a hint of struggle, stress, or uncomfortableness. The hawk seemed to be in deep-freeze mode, where time stops, yet breathing continues.

If a man was treated in this fashion, the man would be in deep distress. His mind would continue to process the various scenarios that might be playing out. His survival hardware would most likely cause him to fight. The man would vocalize, struggle, and maybe even panic. It would be anything but calming, and that would be obvious to outside observers.

Could it be that the hawk is really programmed for catching prey, and depriving that stimulation causes the hawk to be devoid of thought? Or could it be that this man is just naïve enough to think that? I ponder these thoughts and the ramifications of my actions. But does the hawk also ponder? Does it replay scenarios in its mind? *Does it worry about what I am thinking as an observer to this abduction? Could this process be devoid of pain and suffering for the hawk?*

"Hold the rope… Hold the rope… Hey… Hold the rope," Bill signaled to me as I come back to reality from my deep thoughts. Bill proceeded to reset the trap. We are soon back in the tent and I finally eat my sandwich. Over the next few hours, two more Red-tailed Hawks are captured and one of them is kept.

Around mid-afternoon, the flight slowed and the three men felt that they would need to get on the road. It was decided that the apprentice had an appropriate hawk and that the other hawk could be released. The tape was removed from the hawk's feet, the nylon stocking pulled over the bird, and, finally, the hood was untied. The bird instantly rebooted and came out of deep-freeze mode. Within minutes, the bird was back in the air and heading south.

It was time to pack up, and everyone helped load the trucks. When most of the work was completed, the apprentice's hawk was brought out of deep-freeze and placed into a large dog carrier cage. A perch stretched across the cage, and the apprentice placed the hawk's talons on it. It took a few tries before it recognized that was a place to stand, rather than on the floor of the cage.

The apprentice was very excited, and the two falconers reviewed the process he should follow tonight when he gets home. The men offered last bits of advice to the apprentice, assuring him that he was prepared for the upcoming weeks of training with the new bird. The falconers said that they would call him tonight to make sure everything was going according to plan. The falconers also said that the apprentice could always call them day or night for any little issues that might arise.

All three men were supportive of one another, acting professionally as if they were at work, and agreed that the day was a success. They all gave me last minute details about what happens next with the hawk, how to find more information on falconry, and then wished me farewell. I thanked the men for allowing me, an outsider to falconry, into their world, and letting me participate in the day's activities. With a handshake and a genuine smile, I departed their company. But I will always remember the men and birds of this unique experience.

CABIN FEVER

The wind is howling outside. I can hear it through the walls as I lay in my cozy bed. My dreams are oddly infused with driving, hiking, and counting. I am anxious, ready to wake, and ready to travel. I am not completely aware of this semi-conscious drive, but the motivation is there. I want to be outside. I want to watch hawks. And yet, I don't know why.

March has arrived, and with it a reprieve from a hostile winter. The ground has been frozen solid white for about a month and a half, and the air devoid of heat or moisture. Unlike many of the hawks I am longing to see, I did not migrate away from the unforgiving weather.

Modern technology has allowed me to remain comfortable in this zero degree weather. Yet, my mind has been longing for a more archaic existence, one that does not involve modernity. I want to be on the ridges where the wind lives. I do not want to stay forever, but just visit for the day. The environmental conditions are much too harsh to stay overnight, and I would undoubtedly regret such a decision.

"BIRD!": AN EXPLORATION OF HAWKWATCHING

What is the draw to the uninhabitable environment of the cold, snow-covered ridges? I ask this question often, for it is curious that a species that has designed his world to be comfortable tries to get away from it. It seems visceral—an uncontrolled longing to be outside, away from it all. In this respect, humans and hawks seem to be connected. What I am describing is not an emotion, but rather a drive. The difference is subtle, but here, the notion of hunger seems more apt than joy. The former is hidden and intrinsic, while the latter is obvious and often external. I do not get joy from watching hawks; I get that from watching my kids laugh. What hawkwatching seems to provide is satisfaction, clarity, and grounding. All of these terms seem obtuse and poorly defined. I wonder, are these adjectives just phantoms of my imagination? No, they are not! They are as real as hunger, which itself seems difficult to describe.

People who know about my hawkwatching ask if I *like* watching hawks. It seems like such an easy question to answer. The obvious response should be, "Yes." But that is not an accurate response. I liken it to asking a hawk if it likes to migrate. Anthropomorphizing like this is usually a meaningless pastime, but the metaphor seems appropriate. The hawk migrates because that's what hawks do!

Whatever the hawk's intellect, I doubt that it even knows it is migrating. Its machinery and command center (its wings and brain) simply autopilot towards the north or the south, depending on the season. Most likely, the position of the sun with respect to the horizon, as well as the length of the day triggers the hawks to migrate; although, temperature, availability of prey, and other sense perceptions surely play a role.

Birds, such as anseriformes (ducks, geese, and swans) and passerines (perching birds) become restless before they begin their migration, often engaging in maneuvers with their comrades. Geese will take off from a pond, circle around, and land back at the same spot. Mixed flocks of songbirds become energetic and remain active nights before they actually migrate. In a similar fashion, hawkwatchers become restless as both spring and fall approach (although the feeling is more pronounced in spring).

I contemplate making a long drive to a hawkwatch, despite the poor conditions. The chances of seeing hawks today is small, and yet, I want to go. I reason that the five-hour round trip would be a complete waste of time, that the efficiency of such a trip would be low, with few birds being seen for the effort expended. Still, I want to go! I am not happy about the thought of going, but I really want to go! I check the weather again to see if the forecast has changed since I last checked an hour ago. Unsurprisingly, the forecast remains the same. The construction detour, which will impede my travel, also remains unchanged from an hour ago. But, I still want to go!

What is it that makes a man restless enough to expend his resources watching hawks? I have thought about this for some time and have several answers, none of which are satisfying explanations.

The first idea is one based in evolutionary biology. A hawkwatcher may be attempting to gain status by showing his wealth to potential mates. Hawkwatching requires leisure time, money, and resources. This could be attractive to a potential mate, who would be impressed by the hawkwatcher's devotion to such an esoteric activity.

A second line of thought is that we are social creatures, and we long for acceptance into our core group. Hawkwatching helps one to gain status within the hawkwatching community, thereby solidifying the role as a legitimate member.

A third line of reasoning is that being a hawkwatcher equates to being an adventurer. Society is enamored with such adventurous individuals and enjoys hearing about their escapades.

The explanations above all seem to be nullified by the fact that I already have a partner who has never been impressed with my adventures, and with whom I already have children. Additionally, I usually watch hawks alone and rarely speak to others about hawkwatching. Therefore, I am not gaining potential mates, not increasing my social status as a hawkwatcher, nor attracting a dedicated group of followers who hold me in high regard. So maybe it is something else.

The latest idea that I have been contemplating is one of catharsis. Aristotle used the term to describe the purging of negative emotions, often in the context of tragedy. Despite our daily dose of violence we receive each night on the nightly news, it is hard to believe we live in a period where violence is at an all-time low. This has been well-documented by several popular authors, (Steven Pinker, Michael Shermer, Jared Diamond, and Timothy Ferris) and yet, the idea of living in the least violent time seems counterintuitive.

Most of our time living as *Homo sapiens* has been embroiled with fear and stress. We lived anxious lives because of the very real threats from the elements, wildlife, disease, and mostly from each other. It is only in the last dozen or so generations that humanity (or at least some parts of it) were free to live lives devoid of direct violence.

Today, most people are so far removed from where their food comes from, where their trash goes, and the effort needed to produce goods, that in many ways, we are less connected to the biosphere than ever before. Despite living in this modern world, our biology remains unchanged. I wonder if we have built into our genetics an affinity for the primitive, albeit war, conquest, or predation patterns. If this unsubstantiated claim is correct, then we may have a natural disposition for such behavior. Instead of playing out such energy-depleting and dangerous activities, we live vicariously through the hawks, who act as our conquering agents.

Most living species are inextricably connected to other species through a predator-prey relationship. They are, therefore, reflexively (and we presume unconsciously) focused on those that they can eat or be eaten by. *Homo sapiens*, like most other highly-developed social species, are keenly aware of other individuals within their clan. We, however, *seem* to be solely enthralled with other species that are not a threat to us nor directly useful to us. Maybe that is because our science and technology has freed us from having to struggle each day with the basic survival tasks, thereby granting us time to observe artistic, rather than functional, aspects of our environment. Or, maybe the reverse is true—that viewing non-essential artistic entities helped us survive because we could exploit new avenues through engineering. It could also be the case that we were engineered for the primitive struggle of acquiring daily sustenance and that those schemes and tools that made us successful remain in place but are now being applied to a wholly new set of circumstances.

Whatever the reason, it seems clear that many of us are driven towards interactions with nature. It may be that the hawk remains in the struggle for life and that we identify primordially with that struggle. Our current condition may be one of comfort, but we may still recognize the universality of the survival of the fittest. Watching the hawks as they conquer the air, understanding that they must fight to conserve their energy, all while watching out for other birds of prey, potential prey, and newly erected towers, allows us to feel the tension of living in a harsh world.

The catharsis comes from our knowing that nature is both cruel and beautiful, that some of the birds will not make it to their destinations. We are forced to balance the awe we have for these masters of the sky with the knowledge that some will be killed by the elements, other wildlife, disease, and each other. We identify with this struggle, emotionally mourn the loss, and remain hopeful that some of the birds will be here next year. We then go home to our safe, well-stocked, well-protected nests, where we have the luxury of reminiscing about the day's flight, leaving behind the emotional euphoria and stress of what we have seen. Just writing about it tires me out. And yet, I still really just want to go!

ESSAY SIX

BIRD'S EYE VIEW

Alien abductions exist. I know, because I was abducted when I was young. I remember it well. I was very hungry that day. Then, it happened! They covered my head and probed and prodded me. After some time, they let me go. Like others who have been abducted, I think that they put something on me or in me. Maybe to track me. I can't be sure. I see the aliens everywhere, and they see me. I think they watch me. I know they do!

I grew up along the ridges of the Appalachian Mountains, and I return here year after year. These ridges seem to have a magnetic attraction within my mind. I was born here and I have given birth here a couple of times. When I leave, I know I will return. Why, I do not know. These mountains and ridges are just part of me.

My name is Aquila chrysaetos, but everyone calls me Goldie. I am special. I am beautiful. I am strong. I am feared. I travel a lot, but I get recognized no matter where I am, especially by the aliens. They point at me. They look through special glasses to see me. I know, because I see their eyes through those same glasses. They just stop and stare when I come anywhere near. Sometimes, on purpose, I come really close. The aliens get excited. They also seem to like the ridges, especially on sunny, windy days. That is when I see them in the greatest numbers. They gather together, and lift and look through their glasses at the same time. I do not know why they look. But they do look at me.

I like to be away from everyone, especially the aliens. When I am alone, I feel free. I go where I want, when I want. Most of the time, others get out of sight when they know I am coming. It is best that way. I have been known to damage those who cross my path. Sometimes, I surprise animals. I rush over a blind hillside, and then it is over. I am a hunter. I am a killer. I prefer rabbit. That is my favorite flesh. Other varieties will do, but it must be flesh.

I have a reputation. Everyone seems to respect my presence. Everyone except the Corvids. Those damn Corvids! The Jay family is a loud, obnoxious bunch. But it is the Crows and the Ravens that are menacing. They are said to be highly intelligent, but they risk their lives when they harass me. One wrong move and I could end them. That does not stop them though. I hate them. I leave when I see them.

Every year, I see something new, something old, and something bizarre. This year, I see all three simultaneously. On top of the ridges, there are new fan-looking structures. They look like the old ones I have seen before. These are much bigger. They are bizarre in that they move so slowly. I think I will go investigate them. They do not seem dangerous. They look like spinning machines. They are kind of beautiful.

I am getting closer to the wind machines, but what is that ruckus? It appears a group of the Turks are fighting down the hill. I'm going to see what this is all about. Look at them scatter when they see me. Smart move, fellows! Looks like a deer was killed here. Good! I am so hungry. I think that I will tear off a marinated piece of flesh. This makes me happy.

The Turks are clearly not happy that I am here, but they better keep their distance. I have heard that some of my kind have been specially trained to kill small deer. I also heard it was the aliens who trained them. Killing a Turk would be no big deal.

This deer is tender and I think I will tear off another piece. Wait! Something doesn't seem right here. I see marks from the alien transport machines on the ground. Lots of deer get hit by the aliens. This crime scene looks different, though. The tracks in the mud stop at the deer. It looks like the alien transporter backed up here, killed the deer, and then left. Maybe the deer was already dead and they simply dumped it here. Why would they do such a thing? The aliens are wasteful. I will take another bite of the deer while I think. And maybe another.

I can't pinpoint it, but something just seems wrong here. What is that small green glossy box over there on the tree? It has a small, blinking green light.

FLASH!

What was that? It was as if the sun were right in front of me, in that box.

FLASH!

There it was again. I think the aliens put another sun into the box. I would not put it past them. The aliens love boxes, bottles, and packages. They leave them everywhere. Clearly, they do not love this world, probably because they are not from here.

FLASH!

Whatever that box is, it is irritating. It is becoming so annoying that I am thinking about leaving. Whenever I move, I get flashed. The box seems to be triggered by me. It is like it was set up for me, like it was waiting for me to arrive. I don't like this! This feels familiar and wrong. I've got to get out of here. It's a trap!

Thankfully I am able to move away from that deer and that irritating green box. I can certainly cover distance quickly. Look at how far away and safe I am. The Turks go right back to the deer. They are primitive! Look, there is another of my kind coming in the same direction. Looks like a youngster. I wonder if he will be the next victim of that trap. I could go back and try to persuade that fellow to stay away, but I won't. My kind just doesn't have it in us to work together. We are an independent type. We have dignity!

The Jay family, that I mentioned before, they are completely different! Those loud, obnoxious, needy individuals function as a small clan. They vocalize every move of everyone including themselves. If an intruder is in the area, they make a scene. If they find a small bit of food, they let everyone know. If they can't see each other, they squawk. And they work as a gang. They will harass anyone or anything, if there is enough of them. I guess this is admirable in a sense. They just watch out for one another.

Photo by Brian M. Wargo

A Corvid flies away with a fleshy morsel of food.

They are really harmless, just irritating. They are not terrorists, like the Corvids. Oh, how I hate the Corvids. They harass! They injure! They mob! They are baby killers! If I get a hold of one, I will show them who is boss. They are conniving cowards that only attack from the rear. When I try to eat, they will pull and tug on me. When I turn to grab them, they are out of reach. They are as slick as they are sleek. They are bullies!

The spinning machines are close now. I think I will glide right through them. They seem to be in the best place for the moving air. Maybe they move the air. Maybe that is how wind is made. Maybe they are the source of all motion above the land. They are so simple and beautiful. Look how close I am to the end that is moving! Wow, it is moving much, much faster than I thought. Here comes one of the spinning arms now. I wonder what they are made of? Maybe beak material. It is kind of shiny and it is so close to me, almost as if it is going to hit...

Photo by Brian M. Wargo

This juvenile Golden Eagle could be the next to be trapped and tagged. It may also be the next casualty of the wind turbine industry.

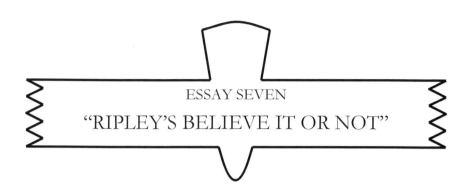

"RIPLEY'S BELIEVE IT OR NOT"

Believe it or not, hawks do not require cliffs and hills to catch a free ride. As long as there is wind, no matter the source, the hawks can exploit it. When land meets water, temperature differentials are bound to exist. This is because water has an enormous capacity to absorb heat. The temperature of water changes very slowly while the temperature of the land changes quickly. For this reason, coastlines usually have a breeze.

During the spring, the water remains cold and the land warms during the day. Cold air is denser than warm air, thereby causing the warm air to be buoyed upwards. An alternate description of this is that warm air rises, although that is descriptive rather than explanatory. At night, the land cools quickly and becomes colder than the water. Again, the denser cold air displaces the warmer, less dense air, now causing the wind to shift out towards the water. Regardless of how you look at it, the cold air moves in from the water during the day, and moves out towards the water at night.

This phenomenon neglects all other forces acting on air. In reality, the earth's spin, humidity, areas of low and high pressure from surrounding areas, and even the beating of insect winds have an effect on the chaotic system that produces wind. While not as simple as first described, it can be safely stated that, where landmasses and large bodies of water meet, wind can be expected.

Wind near the coastline forms convection currents. Think of air as forming big, slightly squared circles that circulate round and round at the border of water and land. During the day, the circles go upwards over the land and come back down over the water. This process reverses at night.

Convection currents do not form over open water because the surface of the water is about the same temperature. Migrating raptors gain nothing from flying over the water, other than increasing their chances of drowning if they get tired. For this reason, soaring birds tend to stay near the shore and ride the updraft. Hawks need this free ride when migrating in order to conserve energy. This lessens their caloric intake while they travel, which is important since every other migrating hawk will also be competing for the limited food sources.

Hawks that can go without eating are at an advantage over those that must also look for prey as they migrate. Since the density of birds on migration routes is staggering, certain prey animals, such as mammals, can quickly become scarce. Hunting itself uses energy, energy that no longer advances the bird along its migration path.

Hawkwatches on the Great Lakes are exemplars of the concepts described above. Hawks that are not on the superhighway of the Appalachian Mountains may head north from a variety of routes from the western side of those mountain ridges. At some point, they will hit the water of the Great Lakes. From there, the vast majority of raptors will fly around the lake rather than over it. These hawks will use the wind currents near the lake to propel them while they frugally expend their reserves.

Situated on the coast of Lake Erie near the Pennsylvania/New York border is Ripley Hawk Watch, which consists of three different watching areas all within a miles of one another. The three sights give a clear view of the lake and are perfectly suited for viewing the type of migration described above.

Ripley Hawk Watch is a productive site because of its topography. An ancient seabed extends inland and forms a small escarpment. This small bump in the otherwise flat area around Lake Erie provides a ridge that the birds can use to gain lift. And, any advantage that can be exploited by the hawks, will be.

Compared with other hawkwatches, Ripley has very high spring numbers. Here, the meeting of water and land discussed above, coupled with the fact that the escarpment funnels the birds into a narrow gap near the hawkwatch site, produces steady flights, especially for Turkey Vultures.

An alternate, less technical, and purely speculative theory has been proposed that may account for the high raptor numbers at Ripley; raptors may simply find joy flying past the dynamo couple of Gil and Jann Randell, who are the backbone of Ripley Hawkwatch. Gil has been working at the site since 2003 and is now the sole counter, compiler, and coordinator. Always beside Gil is his companion, Jann. She takes care of the secretarial duties. This includes maintaining public relations, leading the meet-and-greet department, maintaining the clickers for the Broad-winged Hawks and the Cooper's Hawks, and, most importantly, she chairs the "What can I do for you honey?" department. While Gil is one of the nicest people you will encounter hawkwatching, his status in this category wanes when Jann is present. The reason is simple; Jann is warm, welcoming, fun, charming, and just plain nice to be around. Her laugh is infectious, as is her hospitality. Once you start to speak with Jann, you will hardly notice Gil counting the birds.

If you ever visit Ripley Hawk Watch, Jann will be the person who comes over to you. She will tell you all about the site, the best place to look into the sky to see the hawks, where to sit to remain warm, and will offer you anything you might need to make your time enjoyable. She accomplishes all of these roles while keeping Gil comfortable and focused on the birds. Together, Gil and Jann make an adorable hawkwatching team.

It should be noted that sitting at Ripley Hawk Watch can be a test of endurance, for the cold wind can be difficult to deal with. Even a seemingly mild day inland can be downright blustery at the site. Gil and Jann often sit with a blanket over them, often behind their car, which acts as a windbreak. Unlike some hawkwatchers who freeze to the bone during their day of counting and then take a few days to warm back up to normal operating temperature, Gil and Jann are there every day from March 15 to May 15. This is quite a commitment, but the payoff is sometimes spectacular. In 2014, over a two-day period, 5238 Turkey Vultures passed the site. About a month later, 46 Bald Eagles passed on a single day. The site also had a Broad-winged Hawk day of 6360 back in 2009. These are huge numbers, especially for the spring migration.

Gil is an expert hawkwatcher, but you will never know it from the way he talks. Visitors are welcome to help spot birds, which are usually coming from the west. Less-qualified hawkwatchers or hawkwatchers from more mountainous sights might feel the need to get a quick identification before the bird is out of sight. At Ripley, however, the birds are often spotted at very far distances. This allows a seasoned hawkwatcher, like Gil, to manage several birds or groups of birds in the sky at one time. Here, hawkwatchers can spot hawks coming in on the horizon, take their binoculars off those birds, spend time identifying closer birds, and then go back to the distant birds that are now closer and easier to accurately identify.

The hawks at Ripley Hawk Watch are sometimes very high in the sky and can be difficult to identify. Distinguishing between birds requires patience, and Gil is a master at bird management. He focuses on the hawks and keeps on them until they are identified. Once he identifies the hawks, Gil manages the numbers by using a hand clicker for each species of raptor. He then moves onto the next bird, and then the next bird, and the next bird without break or hesitation. Anyone who has attempted to look through binoculars this long without a break knows how taxing this can be. Gil is like a marathon runner; he just keeps going and going.

Gil is clearly the master at Ripley. All information goes through him. A group of spotters and regulars sporadically attend and help scan the sky. They show great respect to the wise master. This has not made Gil any less approachable, for novices are always welcome. Gil encourages the newbies to work on their identification skills. He knows that his days are finite, and others will need to carry on this important work.

Newcomers may feel that they are helping Gil by offering quick identification of passing raptors. Gil always politely offers thanks for the identification. If time permits, Gil will offer pointers on hawk identification. Sometimes, he has too many birds to manage and cannot instruct you on the error of your identification.

Astute volunteers can judge the accuracy of their identification by watching which of the hand clickers Gil pushes. If they let Gil know that a Bald Eagle just passed, look closely at his right hand and see if he clicks the clicker. If he does, your Bald Eagle is really a Turkey Vulture. Additionally, you can tell if your Cooper's Hawk was really a Cooper's Hawk by listening to the hourly totals. This is when Gil and Jann read the numbers before they reset the clickers for the new hour of counting. If the hour total for Cooper's Hawks is two, and Gil positively identified two Cooper's Hawks and you also positively identified an additional one, then your Cooper's Hawk was not actually a Cooper's Hawk. You should not feel bad though, for you probably did identify a bird—just the wrong type of bird.

Gil Randell is not just a polite fellow and an excellent hawkwatcher, he is also an activist. His report on the wind turbine industry was instrumental in keeping them at bay when they planned on building machines near Lake Erie. Part of his success stems from the fact that Gil is an articulate man and a good writer. This should be no surprise to those who know Gil as Dr. Gil Randell, Ph.D. He earned this terminal degree in literature from the University of New York at Buffalo. As expected, Gil can write! This is evident in the detailed reports posted each day on hawkcount.org. Gil is a stickler for detail in the daily reports and punctually uploads the data at the end of each count session. The care and dedication he puts into his daily reporting carried over to his work detailing the wind turbine issue.

Despite Gil's daily reports, it is difficult to ascertain how unique the Ripley sites are from one another and from other hawkwatches. Anyone who has not visited or driven near the Great Lakes may have a difficult time understanding their effect on the local weather. Thirty miles inland may feel like a different region altogether. In the spring, the weather may be warming, but the lake keeps the surrounding area cool.

All three sites at Ripley have a view of the lake. The site closest to the lake is located in a pullout right off State Road 5. Road noise can be significant, as vehicles travel past at high speeds. On windy days, it is difficult to hear anything other than the howling wind. The lake sits at the edge of a farmer's field, which stands slightly above the water level. The birds fly almost overhead on north wind days.

The most distant site from the lake sits in a vineyard located on a small hill about a mile inland. It is as peaceful as it sounds, with trees to your back and in the distance beyond the grapevines. This site seems the warmest because of the wind-breaking action of the trees at ground level. The third site, called the Barber site, is between the highway site and the vineyard. This site is a wide-open space of interspersed fields and small ponds. I imagine the name originated for the sculpted landscape, like what a barber might do to a field to help hawkwatchers see birds for miles.

Each site has its advantage for viewing hawks, which is usually dependent on the direction of the wind. If the hawks are hugging the lake, then the highway site is best. If the hawks are inland, then the vineyard is the place to sit. When there is a scattering of birds, the Barber site is most productive. Ultimately, the three sites are viewing the same birds, for the sites are within sight of one another. Choosing which site to count from only lessens the strain on the observer's eyes.

In early March, the iced-over lake can be seen from all three sites, but the vineyard site has the highest elevation and, therefore, gives the best view of the vastness of the lake. Small icebergs form when the ice begins to melt and break apart. Letting your imagination wander, you could easily picture polar bears walking across the ice or floating on it. While Gil will keep you abreast to the conditions of the ice in his daily reports, you must see it firsthand to understand this winter wonderland.

Photo by Brian M. Wargo

Ice on Lake Erie can be seen from Ripley Hawk Watch in New York.

I visited Ripley Hawk Watch on April 2, 2015, a day after they posted a daily bird count of 3055 raptors. I figured this might be an April Fool's joke, but it was not. Despite low numbers of other raptors, the Turkey Vultures made a gargantuan push, accounting for over 3018 of the total raptors.

When I arrived with my family for a promising day of hawk counting at the Highway site, we were greeted by Terry Mahoney, who was already starting the early morning count. Terry was huddled beside his truck, sitting in a chair while looking at the sky. He was bundled and trying to sit out of the ferocious wind's path. The snow was gone from the surrounding site, but ice could be seen covering the lake behind Terry. The windswept landscape, the icy lake, and the man dressed for a full day outside made me feel like I was visiting an Inuit camp.

Terry instantly sprang to his feet to greet me as I got out of the car. Terry's eyes were tearing a bit, and I was wondering if something was wrong. Within a few seconds, my eyes were also tearing.

"Little bit of wind today!" I shouted quickly through the wind.

"Yah, just a bit," Terry replied in a more methodical, slower cadence. Within a second or two, Terry eyebrows raised, "I remember you from last year." Terry smiled and we shook hands. We reminisced about the particular days we watched hawks the previous year.

"I was here for the second Golden Eagle in April. I just sent Gil a picture we found of that bird. It was far out, but the picture is not too bad."

Terry replied, "I remember! Gil should be here in just a few minutes. I am just getting us started."

I went back to the car, put on some wind-breaking gear, grabbed a chair, and sat next to Terry. The wind was relentless. Within about 30 seconds, two Turkey Vultures appeared and blew past. These masters of the sky were pointing south, but moving east. They were flying very low, as if they had just taken flight. More began to stream through when Gil and Jann arrive around a quarter to nine.

Gil recognized me despite my bundled appearance, which included a partial facemask. "Brian," he stated as he offered a welcoming handshake, "You made it."

"We all made it. Jeanine and the kids are in the car," I replied. Gil leaned over Terry's truck, saw the other members of my tribe who were in the car next to Terry's, and greeted them with a hand wave.

Terry informed Gil of the vultures he counted as Gil pulled out his chair. As soon as Gil sat down and raised his binoculars, the skies came to life. The birds began to shoot through. Within a short time, others were also joining us, including Julie Leonard, Jill and Berk Adams, Mike Ceci, and, of course, Jann.

The birds increased their rate, and everyone was attentively scanning the sky. Someone in the group asked if there should be a division of labor, where each member would take as section of the sky. Gil responded that he felt too confined by that kind of system.

Jann, helping to contextualize Gil's position, stated, "Gil doesn't like being told he cannot do certain things." The group laughed and made comments about growing up in the '60s, where there was a push against authority.

I suspect Gil felt that certification of the raptors was best accomplished by using the group as a sensing unit and sending all potential birds through his editorial process. This was effective because Gil was clearly the best hawk identifier, and his attention span was virtually unlimited. The group decided that each person was in charge of a clicker, which corresponded with a particular species. I was the clicker for Sharp-shinned Hawks, Jeanine took care of the American Kestrels, Terry for Red-tailed Hawks, Jann for Cooper's Hawks, and so on. Julie offered to take care of the Turkey Vultures, which, at this site, received the most clicks.

Everyone scanned the sky and was welcome to point out each raptor that was seen. After a positive identification, either Gil would tell the watcher of that species to give their clicker a click, or a member of the group might instead ask Gil if they should give their clicker a click. As the barrage of birds began to stream through, summaries of the clicks to be clicked might be given. Gil might state, "Terry, please give three clicks. Brian, would you please count those two as Sharped-shinned? Jann, could you click one Cooper's Hawk? Thank you."

The day progressed in this manner and the birds just kept coming. Bird after bird appeared in the sky. After every identification, Gil would say, "Thank you," for clicking, while Terry might say, "Good spot on that Cooper's Hawk." Everyone encouraged and complemented one another for their effort and their spotting. I began to think that this may be the most cordial group in existence. I imagine if all children were reared in this type of environment, belligerence in the world would become extinct.

Everyone was working hard on this day, but it was Julie that received a physical workout. The vultures were on the move, and it appeared like it was going to be a record day. At one point, it was decided that Julie would exclusively count Turkey Vultures. Of all of the hawks, Turkey Vultures are the easiest to identify with their large black wings, their characteristic rocking during flight, and their tendency to follow one another as if they migrated out of a kettle.

Photo by Brian M. Wargo

A Northern Harrier flies above a Turkey Vulture at Ripley Hawk Watch.

Jeanine helped Julie spot the Turkey Vultures. This was a good job for Jeanine, for she is not really a hawkwatcher. Instead, she photographs birds. Her chronic migraines preclude her from using binoculars. The contrast is too stressful for her eyes. If that is true or not does not really matter to a migraine sufferer; you simply know if something is more likely to cause a migraine. If it does, you avoid it like the plague. Therefore, Jeanine regularly acts as a spotter. On this day, she would simply point out birds naked-eyed that looked like Turkey Vultures, and Julie would verify their identity and count them.

The other observers also spotted Turkey Vultures, but spent most of their time discerning the other non-vulture raptors. Gil was on top of every identification, but was clearly excited to have so many eyes for spotting and directing him to the raptors. For several hours, the pace was non-stop. As someone took a break, someone was there to continue their job.

Around 1:30 p.m., I noticed that I was seeing spots. I also suffer from migraine headaches, and the aberrations in my sight had me worried.

It was about then that Terry said, "I think I have an Osprey over here." But about a second later, he put down his binoculars, rubbed his eyes and recanted, "No, it is just a seagull. My eyes are starting to see things." Julie quickly concurred, complaining that she was starting to see spots. I was relieved. We had spent so much time looking through glass that our eyes and our brains were tiring out.

It was about this time that someone said, "Did anyone notice the clouds ahead of us?" As we looked up into the sky, we all noticed rather threatening storm clouds coming in towards us.

"Wow! Someone want to check the weather on their phone?" As the weather report was read, we all noticed a precipitous drop in the temperature.

Jann stated the obvious, "I think we are going to be shut down here in a few minutes."

Up to this point, we had been enjoying the warmest day of the year. The sun was partially out, and the temperature had warmed all the way to 60 degrees, which was just enough for us to remove a heavy coat or a blanket. Most members also took off their gloves. Everyone was making jokes about the "Florida-like" weather. The reason everyone found the jokes funny was that Ripley and Florida share that coastal feeling, but Ripley's cold is the antithesis of Florida's warmth. Both places are beautiful, but Ripley is a frigid place to watch hawks.

The rain began to fall and everyone began to load chairs into the cars, except for Julie. She was on a mission. Her perseverance resulted in another six Turkey Vultures being counted. The real question is why she was so motivated to count the birds? I think that the answer is that she normally only gets to attend the hawkwatch on the weekends. This day was a Thursday; the weather was warm, and the birds were flying in high numbers. What could be better? Why would anyone want to stop? But, of course, we did!

This five-hour day now stands as Ripley's record for Turkey Vultures at 3483. This is in addition to the other raptors of the day, including: one Rough-legged Hawk, 17 Red-shouldered Hawks, 104 Red-tailed Hawks, 29 American Kestrels, 128 Sharp-shinned Hawks, six Cooper's Hawks, eight Northern Harriers, a Bald Eagle, an Osprey, and a Peregrine Falcon. Combined with the previous day, which now is the third highest day on record for Turkey Vultures, you find 6502 Turkey Vultures passed Ripley Hawkwatch in a mere eleven hours of counting.

This extraordinary day was one of the best overall days of hawk counting in Ripley's history. I feel fortunate to have participated in the count. I understand these type of days are the payoff for the really difficult days, where the weather is miserable, the hawks just are not flying, and the company is either less than optimal or simply non-existent.

Counting hawks is a thankless job. No money is made, just expended. There are no promotions to higher positions, no tax write-offs, no increased status within the local community, no respect from scientific journals, no parity with donating time towards charity, no one honks the horn in support of the hard work, no one stops to shake your hand, no one wears ribbons to signify your dedication and sacrifice, no nothing! Yet, extraordinary people, like the Randells, dedicate their time, expertise, and money in an effort to understand these often overlooked and formerly-persecuted avian species. Gil and Jann spend two months of the year working daily at Ripley

Hawkwatch—they then spend their evenings entering data, researching the weather, and preparing for the next day's count.

I was going to ask the couple why they do it, but, quite honestly, I was afraid to ask. For what if they do not have a sufficiently good answer? What if they analyze the situation rationally, decide that there are social and material payoffs that are far more lucrative than hawk counting. What if they decided to stop? My desire to answer a question could have dire consequences. It just was not worth it.

I know why hawk counting is important. Without it, reliable data about the population of raptors is just theoretical. The Hardy-Weinberg principle in ecology describes equilibrating predator/prey relationships. It supplies a mathematical formula for describing how populations fluctuate over time when coupled into a mutualistic relationship. More rigorous mathematical and scientific models also exist. But, in the end, a physical count must be made to determine the correspondence of the theoretical with the actual. Every business owner knows that inventory must be taken occasionally, for the unforeseen happens. Nature is the queen of the unforeseen. We must measure Nature and her bounty to find out the current state, not the predicted, past, or projected state.

The question that should really be asked is why hawk counting is not held in the same regard as other philanthropic or charitable work? Why do we not have "I Support My Local Hawkwatch" bumper stickers or one that reads, "My Son Is A Hawkwatcher"? I suspect it is the same reason that other forms of science are not popular in culture. It is not flashy, does not produce a quick emotional response, and its fruits are not immediately seen. I know what it takes to do this kind of work, and hawkwatchers everywhere have my admiration. That is small consolation for the arduous and important work they do. Nevertheless, I will say, "Thank You!"

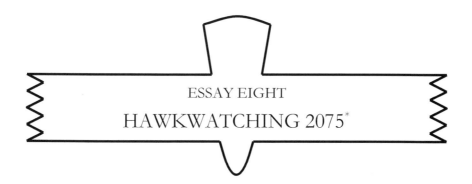

ESSAY EIGHT
HAWKWATCHING 2075*

*This chapter is not indicative of the rest of the book! If you choose this chapter first, please pick another. My father is an engineer and may be the only person who would enjoy this sarcastic non-sequitur.

Welcome to the Natural World Conference, sponsored by eNature®. eNature® – Where Nature seems Natural!® I am your host, Abdel Vichcamer. I am delighted to be here to discuss with everyone the exhilaration we all feel when we think about nature. We have some really great products this year and I hope everyone enjoys this event as much as I surely will.

As I prepared these opening remarks, I thought about what it means to be a *Natural* lover. I was reminded of my grandfather, who used to love actually going outside. He would drive to the hills and hike to a suitable spot, pull out heavy, bulky binoculars, and try his best to stabilize these massive glass filled tubes as he scanned the sky in search of a bird. We have all seen the videos of these early pioneers. Their eyes squinting from the sun as they gave their arms a rest and pulled the binoculars down from their faces. They would do this in all kinds of weather; warm, cold, sunny, humid, dry, windy, you name it, they would stand in it!

These were real men and women! They literally had to go out into the elements and count hawks. At the time, it was the only way to determine their numbers. Some of these early hawkwatchers were really good! They would show up day after day, looking at the sky, often alone. ExtaEyes® was not even a thought then. The parent company, Tinder Live® was still watching people legally. This was before privacy laws began terminating voyeur sites. Thankfully, the animals were never granted human status under the law, so watching animals became a billion dollar industry. I wish I had some of that original stock!

Anyhow, all that technology eventually started popping up all around the country's few remaining pristine locations, where completely wild animals literally roamed free. There was no DirectPather® routes set up to guide animals around urban sprawl, away from transportation routes, and those restricted areas too toxic for them to remain reproductively viable. The animals just went wherever they wanted, whenever they wanted. But, I digress.

Let me get back to the main idea. ExtaEyes® coupled with Animal Identify® technology really started the Naturally Natural movement. Finally, reliable identification of remotely sensed wildlife allowed the viewer from afar to find and see living creatures in nearly natural conditions. Within a few dozen versions, BirdIdentify® was operational. Who would have thought that those pesky birds would be so difficult to identify. I am told that it was not an artificial intelligence (AI) problem, but a physical problem—the distances were too great, the speeds too high, the motion too chaotic. But that is why we are so excited to introduce SkyHawk2100®—the first complete hawk identification system that identifies each hawk by lineage. Unlike earlier versions, which simply identified the SAS (species, age, and sex), the 2100 model traces each bird to their parents, and their parents, and their parents. All the way back, up to 20 generations in some species—a truly remarkable accomplishment!

This is why SkyHawk2100® may be the most important invention since the ISGenocider®. The invasive species genocider has all but eliminated the House Sparrow, the House Finch, and the European Starling from North America. Now if we just bring back all of the species we lost before we got around to making ISGenocider®.

Just kidding! Most of those species were lost from habitat loss and the great biochemical revolution of the 2050's. Oh Boy! I said it! I am sure I will get another demerit point on my stock wealth indicator. Not the first time and surely not the last.

Anyhow, I have taken up enough of your time. It is time to hear from the makers of SkyHawk2100®. Please give a warm welcome to Jim Frank and his design android, Gradient.

Thank you! Again, I am Jim Frank and as all of you already know, this is Gradient. Together, we have put together a great pro-serve. For the old timers out there, the union of products and services industry has reduced material possession by over 30% in the last decade alone. At this rate, we will be material free within the next three decades. Think about the power of this, the complete freedom from personally owning physical objects.

The SkyHawk2100® is the ultimate in the Naturally-Natural movement, fully integrating the experience of the natural world without actually impacting the physical natural world. Every view, every physical sensation is so realistic, so vivid, you have to power down just to realize that you are not actually there. Adventure looms with every experience. The fully automated SkyHawk2100® has the full arsenal of past hawk flights allowing you to simulate past simulations, all while giving the feel of what it was like to collect hawk data a century ago.

For the most adventurous viewers, we now include the option of controlling the AI events by giving the viewer up to 30% input on the outcome of the flight. Of course if you are a purest, you can view real birds in real time using the SkyHawk2100®. That comment always brings the chuckles, doesn't it. Before you laugh, there are a growing number of viewers that actually want to manually control the system. These individuals say they are willing to give up the nearly 99.9999% accuracy to experience the thrill of identifying birds themselves remotely. If this movement picks up any more followers, we are soon going to have to have Grandpa Vichcamer show us how to use those old binoculars. Just kidding. Thanks for joining this viewing and I hope to meet each one of you in your next simulation. Happy Viewing Everyone—Bye, Bye!

ESSAY NINE

URBAN HAWKWATCHING

Hawkwatching in an urban area is similar to visiting any big city, with the preponderance of thought focused on the large number of unfamiliar people. The high density of a city ensures that a wide spectrum of interesting individuals will be present. As the total number increases, statistically speaking, there will be an increasing number of eccentric individuals. Therefore, cities possess greater diversity than rural areas and usually include a motley crew of the downtrodden. One reason for this is that cities have resources and can provide services that rural areas cannot support. The destitute, the homeless, the addicted, and the mentally ill are more likely to survive in a city. People down on their luck who are seeking new opportunities, as well as those who want to disappear may also have a better chance of doing so in the large anonymous crowds.

Cities, unfortunately, are also places where predators know that they can find voiceless victims. Every city has zones where the troubled, the desperate, and the unstable seem to congregate. In these neighborhoods, the rules are different from suburbia and the rest of the city. Those who have made bad decisions, have had bad decisions made for them, or are continuing to make bad decisions, often get stuck in such neighborhoods. After some time this new area becomes home. Children born here are guaranteed to have a more stressful life, thereby shaping their future disposition and potential.

In many cities, these problem areas may be adjacent to the epicenter of the city. Walking just a few blocks may take a person from areas of elegance to areas of abject poverty. Some of the shadiest characters live in between these zones and know how to negotiate the wants and desires of both areas. It is these characters that seem to taint the overwhelming majority of people that are stable and living productive lives. Because it may be difficult to know who is benevolent, most city dwellers assume that others are malevolent.

Making an error with this conservative stance does not make individuals less safe, where the opposite is clearly true. Therefore, urban culture has evolved a more stoic, less friendly disposition in the name of safety. These are realities that every city must deal with and that urban hawkwatchers must also negotiate.

While all this sounds grim, cities may actually be less dangerous *overall* than rural areas. There may be less crime per individual, but since there are so many more people living closely together, there is more crime per acre. Nevertheless, not every acre is equal.

Urban hawkwatching usually has urban decay juxtaposed with some type of natural beauty. These are wild places, not because of their pristine and undeveloped nature, but because of their overdeveloped, already-used-and-abused nature. Real estate in the city is too valuable to leave undeveloped, so those areas that are not currently inhabited are often areas of blight. No-man's-land areas that have been abandoned by industry, ignored by the city, and are problem areas that no mayor wants to address or visit. These areas, however, may be suitable for a hawkwatch. The graffiti, the litter, the filth does not concern the hawks, but keeps most people, vehicles, and industry at bay, thus providing open and somewhat quiet hawk viewing opportunities.

In general, hawks do not live in cities, but just pass overhead. Cooper's Hawks and Peregrine Falcons may occasionally try their hand at city dwelling, but this requires dodging moving vehicles, potentially eating poisoned prey (mice, rats, birds, etc.), and negotiating lots of pane glass windows. The odds are not in the hawks' favor. However, abandoned areas may have weeds and other vegetation growing up through the pavement or making a living from the urban soil of rat feces, decomposing garbage, and exposed dirt under decrepit structures. Hawks flying through the city may decide this is a unique hunting area, and may divert their flight towards this green island in an otherwise concrete forest.

Where the hawks fly, the hawkwatchers will go. These urban wastelands may therefore provide a spot for viewing the migrating hawks because of their expansiveness and lack of activity. These forgotten, unsupervised areas can also be a draw for those with lawless intents and activities. Thrill seekers may visit these areas in order to get that sense of adventure from putting themselves in a less known, less comfortable, maybe less secure setting. This is the same allure many look for when getting out into nature. Some adventurous types seek pristine, nearly inaccessible environments, where wild animals still live. The potential for an encounter with a Black Bear, Coyote, Bobcat, Mountain Lion, Rattlesnake, or a Fisher invokes a sense of adventure. Voluntarily putting oneself in a potential dangerous situation is invigorating. Even visiting areas where wild animals have long been extirpated is often enough to invoke a bit of adrenaline. Just knowing these areas were once a serious hazard to humans is alluring.

Some hawkwatches remain pretty wild. Often these lands are on or near a steep cliff or are simply difficult to access. This makes building infrastructure prohibitively expensive. Whatever the reason, these areas serve as the last bastions of undeveloped forested land. Psychologically, this may be part of the draw for many hawkwatches, who long, cognitively or vestigially, for an earlier, less developed time and space. It could also be that hawkwatchers may be unconsciously searching for a sense of adventure. This may sound nefarious, but next time you are at a hawkwatch, use this litmus test; ask yourself if you would feel comfortable sleeping there overnight. The answer will probably indicate that some hawkwatch areas are wilder than we think. It may be that hawkwatchers get a rush from being on the frontier. Urban hawkwatches may also draw hawkwatchers for their wildness. Here the potential danger is not from wild animals, but from urban decay, rotting infrastructure, and from members of our own species.

Cumberland Gap Hawk Watch, which is a short walk from downtown Cumberland Maryland, fits the above description of an urban hawkwatch. To get there requires driving through the city to the west side of town. On top of the mountain lies an abandoned manufacturing site. A mangled metal gate blocks the private road that leads to an abandoned plastics factory. Vehicles are no longer allowed up the road, so cars must park at the gate near the bottom of the mountain. Occasionally, the locks holding the gate shut are destroyed, allowing for a drive up the very steep, very long, deteriorating road. Most of the time, however, getting up the mountain requires hiking past the gate and up this road.

As you begin walking, you gain elevation quickly. Your calves will undoubtedly burn because your feet can never find a rock to serve as a step. Instead, your ankles must bend to the angle of the road with every stride. The totality of the scene is something out of a movie, with the mangled gate, the long decrepit road, the uninhibited building sitting alone on top of the hill—it can be a bit unnerving.

At nearly the top of the mountain, there is an intersection with the factory entrance and two other dirt roads. Making a left takes you to the abandoned plastics factory. It is gated, but more often than not, the gates are bashed open. From this vantage point, the insides of the factory can be viewed. The doors are open and clearly uninvited guests have entered. Many spray-painted their names on the walls, scattered the remaining inventory in the parking lot, and actually destroyed parts of the building. It is best not to go into this building, for you could be prosecuted for trespassing.

Traveling straight, past the factory, a rocky dirt road takes you up into Wills Mountain State Park. Four-wheel drive is required and most people would not think of subjecting their vehicle to such an abusive road. After traveling a few hundred yards, there is a small pull off is where the unmarked trailhead to the hawkwatch is located.

Starting up the path transforms the urban landscape into a beautiful forested area. Hard rock lies just under the organic leaf litter and pine needles that make up the unique and delicate soil. Unfortunately, where people walk, the soil disappears. Beautiful gray outcroppings are the scars of human travel.

Photo by Jeanine Ging

A long hike is required to reach Cumberland Gap Hawk Watch.

Continuing along the path, you quickly approach cliffs, which are splendid, until you notice the graffiti. Teenagers feel the need to express their love to one another by publically writing their initials inside a misshapen heart. If not in love, they find other symbology to express their current mind set. I hope the image in their mind is more coherent than the jumbled and confusing paint splattering on the rock. Thankfully, the ultraviolet light of the sun fades these bright painted colors to earth tones in just a few years.

Continuing up the side of the cliffs takes you to a rock ledge that jets out of the cliff faces. This is Lover's Leap, also known as the Cumberland Gap Hawk Watch. Here, you find a breathtaking view of the city of Cumberland, and the massive gap between the ledge you are standing on and the mountain on the other side of this substantial rock cut.

Directly below the ledge is a double set of railroad tracks, a stream, Alternate State Route 40, and another set of railroad tracks. This last set of railroad tracks is home of the Western Maryland Railroad, which provides passenger service for tourists. A steam engine pulls passengers up the mountain towards Frostburg, Maryland in the morning, and brings them back in the afternoon. Adjacent to the tracks is the Great Allegheny Passage bike route that connects Pittsburgh, PA to Washington, D.C.

Near the double set of railroad tracks is a large stream. If you follow it southwest into the city, you will arrive at the site of the Chesapeake and Ohio Canal towpath, which is how goods use to make their way across the nation. These canals and natural waterways had small squarish houseboats that functioned as barges. A beast of burden would pull the boat by walking along the path, with a rope connecting the animal to the boat. It was a long, slow, but surprisingly efficient means of moving freight. It survived until the advent of the steam locomotive, which transformed transportation overnight. These vestiges remain in Cumberland and provide a charming picture of a city with so much potential. All of which are easily viewed from Lover's Leap.

Standing on Lover's Leap is a tradition. One inscription carved into the rock lists the year as 1876. The mythology of this spot is that young lovers who would rather not live than have to live apart would consummate their everlasting love by leaping to their deaths. This is probably not true, but dying from this rock is a real threat. A 200-

foot fall would ensure a ticket to the morgue. When the wind is blowing in October, this is a factor that must be considered, for a strong gust could topple an unsuspecting visitor. Looking up provides a panoramic view of the surrounding mountain ridges. A perfect spot for viewing hawks.

Photo by Jeanine Ging

The author views a distant hawk from Lovers Leap.

In the fall, migrating raptors come from behind the ledge, suddenly appearing overhead as they emerge from the tree line. The birds then cross the gap, thereby providing good views. If the wind slows after a windy day, as often happens in the afternoon, the late starting eagles fly across low and slow. The Golden Eagles here are probably the best bird to see, for they majestically and stoically glide across the hawk site in a manner that is always surprising. The sun seems to hit them so that the front edge of their shoulders glows with a golden hue.

Aside from the raptors, there are few visitors to Lover's Leap when the gate is closed. The hike itself is daunting. Those who do make it may or may not be looking for outdoor adventure. Some are clearly out to blow off steam. You can usually hear them coming because people who drink lose the inhibition that regulates voice projection. Unfortunately, there is nowhere to go to get out of their path. Climbing down the cliff is not an option. You just stay in your position and see what happens. With that said, the natural world often relaxes people and allows them to put their guard down. Usually when a group sees what I am doing, they are interested and courteous during their stay on the rock ledge. After their stay, they often wish me luck, offer me a beer, and move along.

Families that stop and chat are usually bonding in the outdoors. They stay a bit longer and try to learn something about hawks and migration. Almost all agree that this spot is awe-inspiring and comment on how nice it is to be out in nature. They then, unknowingly, make an outdoor faux pas or two. They may throw rocks off the ledge, which is a real threat to the rock climbers who frequent these cliffs, or they may break branches off a "small bush" as a nervous habit. What they do not understand is that that bush is really an old tree; one of the few survivors of this wind swept, nearly soilless rock face. They may even use their shoes to dig up the soil and brush it off the rock, not knowing how many decades it takes to generate that soil.

What most visitors have in common is the message that Cumberland is going downhill. It is interesting to hear from *what appear to be* rather rough hooligans that "Some of these people here are real dirtballs." Most claim that Cumberland, in an effort to bring jobs to the area, has attracted four penitentiaries. Along with the jobs came the families who visit the prisons. Instead of making long commutes, they just moved to Cumberland. Some say the new residents created an unsafe feeling in certain areas of the city. Others posit different claims for the cities demise.

If you were to continue along the path from the hawkwatch, the wooded forest extends for many miles. This is wild territory. Part of it is owned by hunting clubs, who regularly harvest bear from this area. One of the hunters let me know that there are some "big ones back here." His son immediately asked if I carried bear spray. I let him know that I did not. He replied, "You better!" I took his advice, although I have yet to see a bear here. Others have told me that Bobcats have also been a problem around the area, with an alert being issued for the Cumberland area. Again, I have not seen a Bobcat, but I would certainly like to from a distance.

Cumberland Gap Hawk Watch serves as an exemplar of birding in or near a city. Because cities are often founded on certain natural features like a geological formation, abundant natural resources, or simply its scenic beauty, thinking about birds near a city is not that far of a stretch. Almost all cities settled because of access to water, which continues to be a major draw for people and birds alike. Raptors congregate to areas of water and some, like Bald Eagles and Osprey, will take food directly from the water. Other raptors will indirectly feed on those animals that rely on the waterways for sustenance.

In Cumberland, the mountains traverse both sides of the city, and a river runs through it. The steep hillsides mean updrafts, and since these ridges start near the Great Lakes, these are part of the superhighway of the sky. This particular set of ridges is not the most heavily used route, but some days are surprisingly good. For example, in November, Golden Eagles fly through in numbers. In 2013, 27 Golden Eagles passed in an eight-hour period. Another example was over 100 Red-tailed Hawks flew past in a four hour period on an early November day in 2014. With that said, I have spent many days without seeing a migrating bird. This of course happens at most hawkwatches.

What Cumberland Gap and many other urban hawkwatches offer is a never-ending supply of people and materials that are transported to and from the city. That means that something is always happening, even when the birds are not flying. Cumberland Gap Hawk Watch has trains continually running below it (with an occasional stream-engine), a bike trail, a two-lane highway, and a small river. A look through the binoculars also allows for good views of the airport, which regularly has glider planes. These engineless planes are towed by a propeller plane and fly much like the vultures—gliding and soaring on the thermals and wind currents near the ridges. While the goal is always to see hawks, Cumberland Gap Hawk Watch offers a plethora of interesting happenings that are best viewed from this unique perch.

Regardless of the numbers of hawks spotted, at the end of the day when hiking downhill back to the car, a little voice in your head asks, *"Do you think that the car is still there? Do you think someone broke into it?"* Living in rural areas does not immunize people from having these thoughts; in fact, it may be more likely to be broken into at a remote site verses a busy site in a city. Irrespective, there is a unique feeling in the city about private property sitting out in public.

As your calves, knees, and quadriceps burn as you descend the hill, an interesting mixture of emotions is felt: A feeling of being "out there," all while being surrounded by the city. A feeling of enjoying nature, while being in a developed zone. A feeling of melancholy about the decay of the city and a renewed hope for reinvigorating it. A feeling that I am glad the day is over and a longing for coming back. Like urban living, hawkwatching in an urban area is complicated. It is not for everyone, but others can't live without it.

ESSAY TEN

ANGLERS OF THE SKY

Hawkwatchers are the fishermen of the sky. Comparing the two activities may at first seem odd, but, with some consideration, parallels begin to emerge. To begin with, fishermen and hawkwatchers are both at the mercy of the animals they seek. Perfect conditions do not necessarily result in numbers. Sometimes, fish are just not going to bite, and hawks are just not going to fly. Each day is like playing the lottery; the hope is that today will be the big day in either abundancy or in glimpsing a rarity. Experienced participants understand the role luck plays with each endeavor. What is known with certainty is that the fish will eat and the hawks will fly; the problem is knowing when and where they will do so.

This does not mean that fishing or hawkwatching are devoid of skill. It just tempers the confidence that each can have on any given day. Both fishermen and hawkwatchers can influence their luck by how they appropriate their time and energy. Environmental factors make more or less conducive environments for both activities. Everything ranging from temperature to barometric pressure has some effect on animal behavior. For example, rain douses flight feathers of birds, rendering them ineffectual. Therefore, most soaring raptors choose not to travel during the rain. Fish, on the other hand, seem to be invigorated by a sprinkling. Choosing when to partake in an activity certainly does influence the outcome. The sarcastic reader is surely thinking, "Thank you, Captain Obvious!"

If we explore the idea of luck and choices a little bit further, we find that things are not so obvious. For example, we feel that several days without impressive numbers must increase the chances of success. This is where our statistically inept brains reveal that we have evolved blank spots when dealing with probabilities. Intuitively, we feel that a coin that has been flipped and lands on tails three times in a row indicates that the next flip is more likely going to end up heads. In fact, the odds remain 50%.

Contrastingly, having several birdless or fishless days does seem to increase our chances of having a big day. But is that true? Is it possible to have several birdless/fishless days in a row without increasing the probability of a high-number day? Let us remember that the probability of flipping ten heads in a row is small, but it is possible. The answer is that birding is not quite the same as flipping coins. Coins can yield two outcomes: either landing on heads or tails. Birds do not really fall into the same category of discretely flying or not flying. Birds have an additional degree of freedom, which is where to fly and how long to fly. A bird may be flying; you may just not know it is flying because of your position. The same holds true for fishing—the fish may be eating, just not what you are offering or where you are offering it.

A better way to think of birding numbers is to think about the average number of birds that are seen. For simplification, think that, in any given August, the daily average is about five birds. It is doubtful that five birds will be seen every day. More likely is a scenario in which most days have few birds and other days have large numbers or are birdless. Add them all together and divide by the number of days and you have the average. Since having several birdless days is less than the average, the following days are more likely to be higher because they are more likely to be average, not

84

because you are due a large bird day. Therefore, a really big bird day is not eminent. If you think about it the other way, it may seem more obvious. Having a 20-bird day does not increase your chances of having a birdless day. Clearly, 20 birds is above the average, so what is expected is having more average days. This idea is called regressing to the mean. Over periods of time, the mean value can be found. You can expect that, with some certainty, the average values will hover around a particular value over long periods of time. So, next time you have a birdless day, think, "Tomorrow or the next day is more likely to be an average day."

The situation with hawk migration is a bit more complicated than what I have described above, for different species migrate at different times and are dependent on different environmental factors. This does not necessarily mean that the numbers get more chaotic. Over longer periods, patterns certainly emerge that help inform us about future migrations. Hawks migrate within a timetable. For example, Broad-winged Hawks tend to move in a large push during the early part of September. No one knows for sure what day they will move, but you can be 99% sure it is in September. You might even be able to say with 75% confidence that it will be during a particular week. What day the birds actually migrate is really up to the birds.

Well, kinda! Birds may not really have a choice either. In their brains is a switch of sorts that tell the birds to migrate. Songbirds, for instance, use the position of the celestial spheres, which includes our sun. When the alignment of the sun's rays hits a particular receptive eye cell at the appropriate angle, or when the interval that the sun is shining per day hits a certain value, the bird begins to migrate. Hawkwatchers that are aware of this may increase their chance of seeing migrating birds by looking into the sky when these natural cues happen. Viewing historical data can help pinpoint when this is likely to occur.

Further complicating matters is that hawks migrating do not equate to the hawks being counted. Hawkwatchers may feel that they are scanning the entire skyline, but, in reality, we only see birds that are pretty close to the site. Think about how small a hawk would look if it was at the end zone of a football field and you were at the other. This is only about 100 yards, or about 1/15 of a mile away. If your binoculars are 10 power, you might therefore, reasonably expect to see the same view about 10 times as far away. This still limits our view to less than a mile away. Clearly, it is difficult and probably close to impossible to see the birds that are far out.

It is also impossible to count every bird. We simply sample them. One hawkwatch may be high one day while another site may be low. This could be dependent on the wind, precipitation, or even the emergence of cicadas. But if enough samples are taken from various sites, a pretty good overall picture emerges about how many hawks are migrating, how they are migrating, and when they are migrating. So, there are certainly many factors that are out of the hawkwatchers' control. With that said, we can mitigate our chances of seeing migrating birds by making smart choices.

One way to increase your odds of seeing a hawk is by choosing an appropriate spot. Fish and birds both are interpreters of the landscape. Mountains and ridges exist both above and below water. Birds can be thought of as swimming through an ocean of air, and fish can be thought of as flying through a large container of water. This may sound far-fetched, but air is liquid in that it fills a given container. The main difference is that water, unlike air, has significant mass and is less able to thwart gravity's pull. Air is not immune to this pull and must obey the same laws of nature as water. The gases that constitute air are about 800 times less dense than liquid water, but gravity pulls on it similarly, yielding similar phenomena, like pressure.

Pressure differences cause movement in both water and air. Fish swim by moving their fins and tails through the viscous water. They can push off the water with their fins, much like people use their hands to push water backwards to propel them forward. Birds can do the same, but they use another trick. As air moves past a wing, a low-pressure area is created above the wing, thus producing lift. Hawks, consequently, seek moving air so as to use the power of wind to remain aloft. Fish, similarly, use obstructions in flowing water to remain motionless in an otherwise fast-moving stream of water. Of course, fish also have their own special system for lift. Air bladders within the fish can be squeezed by internal muscles to change their overall density. Squeezing the air makes the fish denser than the surrounding water, thus causing the fish to sink. Relaxing the muscles causes the fish to rise in the water.

The above techniques utilize pressure differences to achieve nearly free motion. In this respect, fish and birds are masters of exploitation. Hawkwatchers also try to exploit the natural world by choosing landscapes conducive for viewing hawks. Some ridges are absolutely spectacular for seeing birds when there is a particular wind. If that wind, however, is a biannual phenomenon, then sitting at that site day after day will most likely lead to some pretty bleak days. A better strategy is to locate a site that has winds that are more consistent, or to find a site that will yield birds with differing winds (direction, speed, updrafts, etc.). Using these strategies may decrease that outlier day of spectacular numbers, but will yield a higher overall production.

Occasionally, a hawk site in which all of the elements necessary for consistently good hawk counting is found. Waggoner's Gap, a hawkwatch in South Central Pennsylvania, is one such site. West winds bring birds to one side of the slope, while east winds bring birds to the other side of the slope. For this reason, Waggoner's Gap Hawkwatch always seems to have birds, regardless of the wind direction. Other hawk sites have reliable winds that are conducive in one direction. If these winds are the norm, then most days are probably decent. A few sites are fortunate to be at the precipice of converging flyways, thus yielding productive flights most days. In a similar manner, certain streams are mainlines to the breeding grounds. Salmon will traverse particular steams, smelling their way back to their birthland. Here, fishermen know that the salmon will be running and hope to coax them to their fishing spot, and, ultimately, to their dinner table.

In summary, both hawkwatchers and fishermen go out in search of glimpsing nature's most efficient solutions for traversing the air and water. Technology has allowed us to travel in both mediums, but we seem to be drawn to witnessing nature's solutions. Some people are interested in the capture, either by catching fish or hawks (falconers), while others are content by just passively observing (hawkwatchers). Regardless, both are rolling the dice each time they go out, not knowing if they will be successful. Both hope that today will be the big day, despite the odds not being in their favor. But if they just keep at it, then success is right around the corner. Or is it?

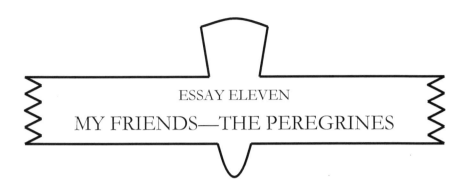

ESSAY ELEVEN

MY FRIENDS—THE PEREGRINES

Early season hawkwatching is always frustrating. Hawk counters are ready with their clickers but, in reality, the birds are just not ready to move. Local hawks tease the hawkwatchers, acting as if they are going south, but then suddenly turn around and head in the direction they came.

The total number of birds counted each day in August remains low; however, some sites will begin reporting early migrators, such as Ospreys and Bald Eagles, by the half dozen in mid-August. Seeing these birds is always a treat and makes the day worth the effort. For most sites, a total count of four birds may be typical, and that is for all species added together.

So, why start so early? I have asked myself that question many times. It is the same question that many who love a particular hobby, sport, or lifestyle confront. It seems illogical to partake in such activities when the odds of having a near-peak performance day are dismally small.

It may have something to do with our psychology, especially when connected to something we cherish. Our minds seem to rewrite history. We remember the high-count days and seem to forget the birdless days. Everyone does it. Think of the dog owner who seems confounded when their dog barks at you, even though their dog barks at you every time you encounter the animal. They may even say, "That's strange! My dog never barks at anyone." What is really strange is that they believe it!

For whatever reason, our brains put more weight on the day we see 100 Red-tailed Hawks and less weight on the other ten near-birdless days. It is as if we create a false history of our lives and then long to recapture past glory, both in birding and in life.

Hawkwatchers are not the only ones that get anxious about the impending migration season. Birds also become restless as the migration approaches. Flocking birds, like ducks and blackbirds, increase their flight activities. Ducks and geese will take off, fly a few large radius circles, and land in the same spot they were originally floating. Songbirds become more vocal and active at night as if they are checking to see if their neighbors are still there.

Raptors are a bit different from other birds when it comes to migration. These birds of prey are not as social as other migrators, so they are not interested in being around other raptors. There is good reason for this; raptors hunt and, therefore, compete with each other for food. The last thing a raptor wants after a successful hunt is other raptors. For this reason, most raptors go through life either alone or with another family member. When they decide to migrate, they do so individually. If they flock, it is usually opportunistically.

Raptors watch other raptors, using them as markers for where the updrafts are located. When one catches a thermal, raptors from all directions will converge to that particular spot. They are not really looking for each other's company, just the same free ride that the other raptor is getting.

Non-raptor migrants also watch for hawks during migration. They are in the open without hiding spots, without their camouflage, without brush to retreat into. Quickly, they can find themselves in danger. This is partly why many species of songbirds migrate at night, when raptors are roosting and not hunting. Another strategy of lessening their chances of being the bird that is killed, is being one of many birds in the sky at any time. If there is a predator and you are the only bird in the sky, you are the target!

Raptors have the luxury of not needing to think defensively. They are not worried about becoming a meal and they boldly migrate during the day. Regardless of their seemingly confident psychological state, raptors act peculiar near the beginning of migration. It is as if they are not sure if they are migrating. A raptor may think, "I think I will just explore the area slightly south of this current position. Or not!"

I am, of course, anthropomorphizing and have no idea what, if any, conscious ideas stream through the brain of raptors. I suspect little actual thought occurs. Many raptors, when not nesting, are kind of always migrating. They wake up, notice they are hungry, jump off the perch, and search for food. If the food is moving south, they are necessarily moving south.

Photo by Brian M. Wargo

A Peregrine Falcon playfully flies near the cliffs of Cumberland Gap.

Hawkwatchers tend to get hung up on birds that are noncommittal about migration instead of enjoying the views of these birds. They crankily and disappointedly state, "Well, that bird is just a local." This may be the wrong attitude, for I have had memorable days hawkwatching when the birds were not yet migrating. For example, there was the day that I witnessed a Peregrine Falcon fight with a Northern Goshawk. The following excerpt is what I entered into hawkcount entry for the day's raptor observations:

At 12:30 p.m., I heard the [resident] Peregrine Falcons acting up. Within a few minutes the agitation of the birds was very loud with a new sound I had not heard before (a short "chuck"). It appeared that after days of deliberation, I finally saw the juvenile Peregrine Falcon that I suspected must be around the two blue steel colored adults. I then saw for the first time a brownish plumage which I had not seen the previous days. Then the adult Peregrine and this juvenile began to lock talons in what appeared to be the most aggressive play I had ever seen. It was then that I realized that the juvenile was much bigger than the adult and had a tail that would expand like a Buteo. I had seen the Peregrine[s] so many times here that I was not using my binoculars. When I finally did grab them, I saw the juvenile was indeed a juvenile, a juvenile Northern Goshawk. For the next ten minutes, I witnessed one of the most aggressive fights of my life. The Peregrine swooped again and again at the Goshawk, who never retreated but kept coming towards the Peregrine. Several times they both landed on the cliffs, taking turns going after one another. The Peregrine has the speed advantage and could gain altitude quickly, then proceeded to dive onto the Goshawk. The Peregrine at times looked like an aggressive Humming Bird, making long pronounce dives. The Goshawk was never deterred; it just kept coming back towards the Peregrine. Finally after minutes of battling, with continuous howling from the Peregrine, and the shorter, muted, "chuck" of the Goshawk, the Peregrine drove off the Goshawk, staying above it as it gave a slow chase down towards the stream and around the bend. [The] Peregrines are regulars at Cumberland Gap and I have seen them dominate every other bird up to this point (Red-tailed Hawks, Broad-wing Hawks, Bald Eagles, etc.), but this is the first time I had ever seen a real fight. This Goshawk was the most aggressive, toughest bird I have seen.

The point is that hawkwatching in late August may allow for encounters that you might not otherwise get when the raptors are serious about migrating. The lazy days of summer may, in fact, yield lazy raptors that are just hanging out, slowly soaring around without rush, without a desire to get as high up as possible in the sky. This often means getting better views of each bird and maybe even a second view if they turn around. And, because they are acting peculiar, they might do something that they normally wouldn't, like land near a hawk site and stay for a while.

This is what I think happened with a pair of Peregrines at my hawkwatch. I suspect that they were considering migrating, but thought, "This is a pretty nice ledge—let's hang out here!" They seemed content sitting on the edge of the cliffs each day, and I looked forward to seeing them every afternoon.

As I sat in the burning sun in late summer, I was often alone. It was nice to have the Peregrines, who were like friends at the site. Since so few birds were moving, the days were long and often devoid of excitement. That is until a Turkey Vulture would accidently fly too close to the Peregrines, who would use the opportunity to stretch their wings. The Turkey Vulture would immediately leave the area, but the Peregrines were excited and ready to put on a show. They would fly high, tumble down, and head right for the cliff face, pulling up at the last second. Their acrobatic display was just beautiful, and they seemed to always give me a close flyover. Other times, when they were not flying, they would break the monotonous silence of the day by calling to one another, although I always felt they were also calling to me out of sympathy.

Sometimes, one of the Peregrines would just pick a nice perch on a rock and sit for hours. It was just the Peregrines and me at the hawk site all day. Usually, I would set the spotting scope on one of the Peregrines, so if nothing was happening in the sky that hour (or the entire day), I could always see a bird.

During one rather dull day, I noticed that I was not the only one scanning the sky intently. Through the scope, I could see the

Peregrine observing each bird in the sky (non-migrating Black Vultures and Turkey Vultures were ubiquitous in August). The big, black eye would intently scan each bird. Suddenly, I saw the eye look up high, as if looking into the sun. I was impressed by its ability to handle the brightness of the day. The Peregrine was carefully studying a bird that I could not find due to the sun's position.

Photo by Brian M. Wargo

An inverted Red-tailed Hawk prepares for its encounter with a diving Peregrine Falcon at Cumberland Gap Hawk Watch.

The wings began to perk up and, in just a few seconds, the bird was clearly amped. Jumping off the cliff, the Peregrine began to pump its wings. In no time at all, the Peregrine was high in the sky and heading for an interesting looking bird. It had the shape of a Red-tailed Hawk, but it was ghostly white in the sun.

The Peregrine engaged the bird and it responded by getting defensive. The white-headed bird flipped over and revealed its pinkish-red tail, and a white-splotched brown top side. It looked like a Krider's Hawk, or an intergrade of one. This would be a rare sighting, for Krider's Hawks are found in the Great Plains of the United States.

Thankfully, I had my camera that day, and I captured the encounter. The "Krider's Hawk" was totally out matched and knew it. The Peregrine would dive on the hawk so quickly, reach its apex near the bird, continue a beautiful arc until it reached the high in the sky, and then dive again. The Krider's did not waste time; it quickly stooped and made for the hills. Regardless, if it were not for the watching eye of the Peregrine, I may have missed this uniquely light-morphed bird.

Photo by Brian M. Wargo

The ghostly-white face and light underside made this Red-tailed Hawk appear like a Krider's Hawk.

I was able to snap a picture of the hawk as it jetted past. Analyzing the shot revealed it was not a Krider's after all, just a light-morphed Red-tailed Hawk. It was an exciting encounter, regardless of the hawk's race. It made a boring August day one that I would not trade for a high-count bird day.

Sometimes, hawkwatchers forget that migrating is serious business. Migrating hawks move with intent. They are more goal-oriented: getting south, finding nesting material, moving from impending inclement weather, finding new food sources, etc. If their fellow raptors, like Mississippi Kites or Broad-winged Hawks, are migrating together, then they are in a race with one another. There is no time for showing off for the hawkwatchers. They often find a thermal, get high, move quickly, and click off the miles. This allows hawkwatchers a greater quantity of hawk encounters, but not necessarily the best quality encounters with the raptors.

Photo by Brian M. Wargo

A light-morphed Red-tailed Hawk stoops in an attempt
to leave the Peregrine Falcons' territory.

August is a slow time for both the hawkwatcher and the migrating raptor, but that does not mean that it is not worth the time and effort. Almost certainly, the numbers will be low, but they may be more meaningful than they appear. Sometimes, seeing just a couple of birds is fulfilling. If you don't believe me, just ask my friends, the Peregrines.

SAFARI IN THE SKY

When people think of the ocean, they think of beaches, coral reefs, and colorful fish swimming past snorkelers. Similarly, forests are thought of as places teaming with wild carnivores like bear and large herbivories such as elk. The sky, however, is thought of as a huge, open, lifeless expanse. Our conceptions of life in the water, on land, and in the sky may need a second look.

It may be surprising, but most of the ocean is devoid of extravagant aquatic life. In fact, most of the ocean does not contain fish. The vast majority is a water-based soup of microorganisms mixed with an ever-increasing amount of ground-up plastic. Only certain regions, where mineral and nutrient rich water are available, does much life exist.

Forests also appear to be an ideal mecca for wildlife; however, if you have ever visited a forest, you rarely see large animals. Seeing a deer is exciting, a moose even more so, and glimpsing a bear is a surprisingly rare occurrence. Many of the predators that are supposed to be in the forest are not just rarely seen, but they are almost never seen. These include Bobcats, Mountain Lions, Wolverines, Martins, and Fishers. Some outdoorsmen go decades without seeing even one of the above predators.

This is why the sky can be such an enigma, for it can be teaming with life. During peak migration, the sky can be filled with huge quantities of top predators. Eagles, falcons, hawks, ospreys, and harriers can be seen passing leisurely by, sometimes all in one day.

So how is it possible that the skies seemingly support big game more than the water or the land? The answer lies in where the organic chemicals are, or more importantly, where they are locked up. Like the rock of the land and the water of the ocean, the air of the sky itself is devoid of nutrition.

Photo by Brian M. Wargo

Ospreys are large predators of the sky!

Birds do not eat air, fish do not eat water, and plants do not eat rock, but all depend on these inorganic substances to support life. These media provide a home for life by supporting creatures that contain the essential chemicals necessary for top predators. The way these organic chemicals are cycled through the ecosystem determines how much life each medium can support. At first glance, forests seem to have the advantage with their huge reserves of photosynthetic potential; however, forests have so much of their carbon mass tied up in wood and leaves that the currency of the forest is low. Oceans can be similarly barren, despite the large area of sunlight and water. The problem here is limited nutrients are just not always available; thus, most of the oceans are fairly lifeless in terms of larger aquatic life. Only in regions where sunlight and nutrients come together does life flourish.

Contrastingly, prairies exemplify a booming economy of life. Their carbon molecules are available in thin, digestible forms, such as grasses and flowering plants. The Serengeti grasslands, for example, produce a renewable supply of food each year, supporting large herds of herbivores and a corresponding number of carnivorous. As plants are eaten, digested, and excreted, the landscape recycles the rare nutrients. The animals themselves also provide reservoirs for these nutrients and continue to spread them through their scat and flesh when they meet their demise.

The sky lacks a photosynthetic base that both the land and ocean have, but the sky does have caloric content. Biomass can be airborne and may move up the food chain. Dragonflies are a good example. They eat on the land but live part-time in the air. Broad-winged Hawks, along with other birds, eat the dragonfly on the wing. Peregrine Falcons also find their food in the air in the form of other birds.

Some migrating birds simply go without eating during migration, thereby making entire caloric constraint a moot point. For the hawks that are looking for a meal during migration, certain routes and areas are more favorable. Fields, like the prairies, are places where life forms flourish and, therefore, serve as magnets for migrating birds.

Putting all of these disparate facts together, we find that ecosystems are not always as they appear. Oceans can be barren, forests may not support as many large predators as we think, and the sky can be filled with animals. Understanding this fact could help people make different choices. For example, outdoorsmen looking for excitement often think of hunting or fishing. Many do not really want or need to harvest the animal, but just want to catch a glimpse of it. This is why many hunters do not actually take the shot, and many fishermen practice catch and release. The enjoyment is in seeing the animals. Cleaning a fish or gutting a deer is a chore that few enjoy. Knowing that such a chore is the end product can actually deter future engagement in the activity.

Hawkwatching may be a viable alternative for outdoorsmen who are tired of waiting around for top predators that are nearly impossible to find, making more work for themselves when they are successful, or tired of feeling that they must actively change the balance of nature instead of just observing it. Hawkwatching does not require more time and effort when you are successful. You just look up and enjoy the top predators of the sky.

If you want something to show, just snap a picture. If you want to feel more active than passive, then become a hawk counter. If you want to find more exotics, try for less common species. Whatever your angle, hawkwatching may have something to offer. All that is required is for you to look up!

ESSAY THIRTEEN
NATURALISM AS A REPLACEMENT FOR ORGANIZED RELIGION

Free thinkers are infamous for avoiding organized religion. This can be a real bummer since religion brings people together and builds community as no other social system has in history. Football comes close, but does not quite have the draw. A commonality with both endeavors is the encouragement of individuals to pick sides or teams. In football, many, or at least some fans, see through the silliness of having to cheer for the proximate team rather than a team consciously chosen due to merit, personality, or showmanship. Religion does not offer this perspective as readily. Blind adherence to the home team or ostracism (or death, depending on the country in which you live) are sometimes the only choices.

The underlying problem lies with having to pick a team, which instantly defines those not on your team as "other." This is not a problem, so long as your team is currently winning. This division of "them versus us" can foster intolerance, which can undermine a free and just society. This discriminatory process is so natural and so comfortable that a self-reflective person might recognize the dangers of such thinking.

The good news is that this is not the only option. Alternate positions include having no team, being a member of all the teams, or supporting those teams that seem to be working cohesively, playing fairly, and consistently.

These last options seem to go against our tribalistic instincts. We evolved with a built-in prejudice for those who share our genes. Biologically, we are trying to get our genetics into the next generation. The thinking is, if not *my* genetics, then at least *my* brother's or *my* cousin's genetics. This makes a lot of sense for species that are non-cultural and lack control of technology and their biological destiny.

For human beings, who have socially conquered Earth's natural environment, this idea is problematic. This biological tenant seeds nationalism and is one of the root causes of racisms and ethnocentrism. Modern humans should judge one another on merit, on how we treat one another and the environment, and how our ideals help humanity and all other living species.

One of the great powers of religion is for everyone on a particular team to check their egos at the door or altar of the church, synagogue, or temple. The acknowledgement that there is something bigger out there, recalibrates our perspective, if only temporarily. Recognizing that our measly existence is puny compared to other entities is healthy. The problem lies in the entity being a deity who has written a book. Taking arbitrary text (which, technically, was written by man, usually hundreds or thousands of years ago), treating it as true, good, and sacred, and enacting those ideas faithfully can lead to serious confusion and contention. The Middle East is a good example of what can go wrong, especially when these ideas mix with political ideas.

Now, imagine if everyone was on the same team. Call it "Team Nature." Everyone is member, everyone is part of it, and everyone understands that each individual is temporary, fragile, and insignificant compared to the rest of the natural world. Those that feel above the natural world can test their strength by moving towards a tornado, standing outdoors for 48 hours in the hot sun of an Arizona summer or the bone-chilling temperatures of a Minnesota night, or not getting vaccinated and hanging out with the few remaining polio-infected individuals in the world.

Most recognize our fragility without such tests, so the majority of the world understands this starting point. This idea, that we are not powerful compared to nature, allows for the almost universal feeling of awe that humans experience when placed in the natural world. By this, I mean outside the grasp of our technological safety net. Taking a hike in the mountains, going outside at night to observe the stars, or sliding down a snow-covered slope all remind us that our current style of living is not natural but is subsidized by a myriad of gadgets.

I want to stress that I am not advocating going gadgetless. My life is absolutely positively better because of hot water heaters, screws, electricity, computers, aluminum foil, soap, etc. However, grounding ourselves back to what is provided naturally—that chaotic, lowest energy level state—makes us appreciate the foundations of life, like the Sun and the radiant energy that allows the dynamic Earth to flourish.

Life, as a shining example of harnessing energy and making the Earth the special place that it is, can be heralded as an example of the wonder of existing. And, life is not a foregone conclusion. It is special in terms of time and place. The Earth was lifeless long ago and will become lifeless in the future. Other planets are either too far or too close to the Sun to be viable candidates for harboring life as we know it.

The above paragraph meets two of the criteria for religion—sacred time and sacred space—Earth and now. Thinking of Earth as a religious symbol should not be much of a stretch, for we use natural examples as our symbols of our most cherished ideals. Countries, sports teams, schools, rock bands, companies, etc. use animals as their symbol or trademark. Even plants and natural formations get notoriety (think Apple or Glacier). We invoke the word natural in an almost spiritual sense—"Our product is better because it is natural."

Let's just assume that everyone can get on board with heralding the Earth as a really special place in our hearts and minds. Let's make it sacred! Now imagine how careful everyone would be with the sacred object during our tenure. We could even pick a sacred symbol. But what would it be?

Life has evolved on the Earth in a myriad of different ways. One of the most interesting is that of life in flight. Insects fly, but using an insect as a sacred symbol will probably cause most people to stop reading this essay (if they have not already). Only raptors seem to majestically soar. They symbolize rising above, conquering gravity, and having complete freedom. Watching hawks as they migrate could be a religious right of passage. Sacred stories could be passed down, creating a mythology that humans love. Stories of a selfless act could end with, "As the man carried the injured friend up the hill, a Golden Eagle flew past, indicating the natural world was watching."

Priests of the airborne creatures could be used for offering blessings. Falconers could ceremoniously pay homage to sporting events by allowing a Peregrine to swoop down the sporting field, thereby inspiring everyone to run their fastest. Sacred stories could be created, such as the marriage of wind and flight. The harmonizing of both the living and nonliving world through flight could be used as inspirational text that values engineering and creativity. Virtues could arise, where nature harnesses nature and is held as enlightened thought.

Ultimately, religions are human constructions that seem to fill the need for belief that is so ingrained in our evolved cultures. Most religions are human-centric, holding that humans are the ultimate symbol of spiritual being and the only creatures with a soul. This often leads to one-upmanship, where humans need to be better than the next human. We like to compete, even with our religions, and try to figure out ways that our religion makes us better than the next group. The consequences are sometimes deadly.

Let's choose to be less egocentric and select non-humans for our spiritual sustenance. Let us choose the raptors, for they are natural symbols of strength, independence, and beauty. Let's also never forget that our choice of the sacred was arbitrary, was a human decision, and was meant to feed the natural hunger for inspiration, mythology, and wonder.

Raptors are beautifully adapted creatures that evolved to fill an open niche in the natural world. Their solution, that of eating carrion, not only allows for their way of life, but also helps to support many others by recycling important nutrients back into the biosphere.

We, too, are but one solution the natural world has concocted; although, nature may want to reconsider, for we may be the ultimate destroyers. Our insatiable energy appetite, our ability to create long-lived toxins, and our production of everlasting and indigestible plastics and waste products that literally choke the rest of the fauna to death seems like an evolutionary mistake.

Photo by Brian M. Wargo

A young Bald Eagle serves as a symbol of Nature.

We need to recognize that we are part of nature, and for nature to exist eternally, we must become better stewards. At one time, we thought that nature would go on with or without us, but now we might have the power to make that choice. At our current rate, half of the species now living may become extinct in the next century or two. Our lives are not just our own; other living creatures must also live by the decisions we make.

Our time on Earth may be ephemeral, yet consequential. For this reason, our stay must be considered sacred, for our power to change the natural world seems almost omnipotent. Let us embrace being part of this biosphere, not the ruler of it. Let us consider nature sacred and respect the process that brought us into existence and that keeps us still.

Let us make certain acts sacrosanct, like observing Nature's solutions with awe and respect. One manifestation of this parable could be in watching the hawks fly. As we do so, we can humbly acknowledge that we have been granted the opportunity by Nature, through chance and the selection process, to engage in this miraculous event called life.

As our fellow humans stand near, we can acknowledge that no one was born more sacred than the watcher next to them. We are all just watchers, finding our own meaning, our own purpose, our own destiny in the majestic flight. As the birds fly by, let us look upwards, give thanks to Nature, and pray to the birds of prey.

ESSAY FOURTEEN
VULTURES GET NO RESPECT

Vultures are notorious for endlessly moving back and forth throughout the day. It may seem like they are migrating, only to return two hours later. Because of this behavior, many hawkwatches have special counting rules for vultures. Some may ignore all vulture activity until a particular calendar day. Others want to see intent, trying to ensure that the birds are not locals out for a day trip.

Even when they are migrating, Turkey Vultures garner the least amount of bragging rights. Most hawkwatchers will qualify the day's total with the number of Turkey Vultures that the count contains. "I had 239 birds today. But 197 were Turkey Vultures." The typical response is always supportive: "Well, at least you saw something today." Or, "Those eagle numbers were really good."

It seems unjustified to dismiss these truly amazing birds of the sky. In many ways, they are the cleaning crew for the country. They dispose of the most putrid materials littering the landscape. Unlike the innocuous discarded beer can, the smell emanating from a dead possum or carp can make a grown man gag. For the vulture, this is the sweet smell of margination—a signal that, not only is there a meal, but one that it is so tender and with that melt-in-your-mouth character that human chefs try to emulate (minus the decay, of course).

Needless to say that a human eating such fermenting effluent would undoubtedly vomit at first, and then leak profusely from his or her large intestine, followed with the possibility of death. Both humans and vultures have their feud with the bacteria that are attempting to claim the spoiling flesh, poisoning it so that they have it all to themselves. It is a pretty good technique, and it works for most animals. The vultures, however, seem completely immune to the infectious agents and the toxins they produce. Their stomach acid is very strong, and the bacteria do not have a chance to survive.

The strength of the acid does have a downfall, though, for when the vultures eat lead, it dissolves completely, thereby entering the blood supply. Lead shot and bullets are the main sources of lead poisoning, usually found in the discarded viscera or animal parts left behind by hunters. Anytime an animal is shot and not found, there is a good chance that the lead will end up inside a vulture.

Lead poisoning is a serious issue with all vultures and was nearly responsible for wiping out the California Condor. The banning of lead in bullets is also problematic, for the alternatives are much more expensive and can be more deadly for those who wear bulletproof vests. Lead is soft and malleable, where steel and brass hold their shape and are, therefore, considered "Cop Killers."

Vultures are not picky about their bullets, and, in fact, are not picky about anything they eat. Black Vultures in Florida target the natural and synthetic rubber of car windshield wipers and seals. If you visit the Everglades, be warned that the rubber of northern cars is a special delicacy for the Black Vulture, for it is salted. Within seconds of walking from your vehicle, the Black Vultures will be inspecting your car. Anything that seems remotely possible to eat, they will try.

This is not just for the Black Vulture, for the California Condor finds trinkets of all sizes, shapes, and colors, and feeds them to their young. This is a leading cause of infant mortality in the condor community. Dissection, radiographs, or natural decay of dead young reveals that their bellies are so full of trash that there is no room for sustenance. Even today, most wild California Condors must be given lead-free and trash-free food from conservationists, in order to avoid a repeat of their previous condition.

Photo by Brian M. Wargo

Black Vultures are ubiquitous at Cumberland Gap Hawk Watch.

The vultures' intestinal fortitude is not the only amazing ability of these creatures. Their incredibly low weight and their large wing surface area allow them to kite, no matter how feeble the breeze. They seem to float in the wind, only rarely flapping their wings.

Unlike other raptors that dedicate muscle mass to flapping flight, vultures seem to emulate marathon runners, who rarely have big arms and pectoral muscles. All that extra mass would need to be carried each mile run, all while contributing little in terms of actual function. For the vulture, it is better to allocate that type of muscle mass to the neck region, so as to be able to tear flesh from a carcass. But again, if the flesh easily peals from the bone, that added muscle mass is just a burden that must be carried.

The basic design of a vulture is that of a flying stomach. No frills attached; just pure function. Very little has been dedicated to aesthetics for this bird. No flashy wing feathers, no head crest or crown, no melodious song, no great intelligence, nothing but efficiency for soaring.

Photo by Brian M. Wargo

Three California Condors soar at Pinnacles National Monument, CA.

Even the brain allocates only what it needs, which, in the case of a vulture, is finding food. A Turkey Vulture's brain dedicates a significant portion to the detection of smell, for that is the radar system that pinpoints the source of death. This apparatus is so sensitive that, during flight, the vulture can rock side to side and delineate between the gradient of odor, thereby giving direction to its origin.

The above description paints a picture of efficiency. But, if that is true, why do vultures spend so much time flying around instead of eating? This is an interesting question, because very few people actually see Turkey Vultures eat. Unlike other bold scavengers, Turkey Vultures are rarely seen feasting on the carcasses of animals hit by cars.

Crows are notorious for eating roadkill, but for these birds, they are probably hungrier. Unlike vultures, crows flap their wings nearly continuously when flying, and also spend large amounts of energy socializing. A daily ritual of a crow entails flying like a maniac to both impress and harass one another. In contrast, vultures expend almost no energy flying. They can, therefore, explore their territory, keeping track of the condition of multiple carrion sites, and scheduling the optimal arrival times.

When vultures decide to eat, they tend to gorge themselves. This allows them to survive for a significant time without eating again. This helps when everything else in the world is going smoothly and nothing is dying. Black Vultures are slightly more aggressive and will catch live prey. This seems to fit their lifestyle, for they tend to be higher strung, including flapping their wings more often than Turkey Vultures. They also do not seem to have the Turkey Vulture's ability to smell. The Black Vulture's radar is in watching other birds that can detect carrion.

Black Vultures and Turkey Vultures garner very little interest from most people. On the opposite side of the popularity spectrum is the California Condor. These sought after behemoths almost never flap. Instead, they wait for the perfect wind. If the wind does not cooperate, they will just stay put.

Condors get much more sympathy than the other vultures, probably because we almost extirpated them. Thankfully, the capture and artificial breeding of the last remaining wild condors worked. Although still fed by humans, dozens of condors now live in the wild, and birders plan trips to see them. One summer, I drove from Pittsburgh to California to view these large vultures. It took some time, but I finally found them flying with a group of Turkey Vultures.

They are majestic kings of the sky. When I first spotted them, I waited for trumpets to sound. Of course, that did not happen, but I did get to observe them for over three hours. Unfortunately, my view was constantly inhibited by the annoying Turkey Vultures that would fly in front of the condors. Every picture I attempted to take was also photobombed by the omnipresent Turkey Vulture.

I began this essay describing how Turkey Vultures are incredible gliders, amazing digesters, and the ultimate in energy efficiency. I made the case for why they deserve our admiration. All of that was fine until their ugly faces interfered with my pictures of their more popular cousins. Like many others, I got upset when a Turkey Vulture appeared in photos intended to show the beauty of another bird.

Photo by Brian M. Wargo

A mature Turkey Vulture flies at Cumberland Gap.

As I write this, I try to recall all of the pictures of Turkey Vultures I have intentionally taken. I quickly realize that I never photographed a single Turkey Vulture for the sake of having a picture of a Turkey Vulture. I have included them in pictures to show the relative size of other birds, how kettles form, or the sheer number of birds in the sky, but never as a portrait.

I suspect they want to have their picture taken, for they fly close overhead and look at me when I am holding the camera. They slow down, stare directly at the lens, and seem to stop rocking momentarily, as if they know this cuts down on the blurring effect. And yet, I will not take their picture! They will even come back again over the same spot and pose even better than the time before, and again, I will not take their picture.

Despite my writing about the vultures' important placement in the grand environmental scheme, I hypocritically ignore them when hawkwatching. But, I guess that is what separates us from the vultures—the aptitude to know better, and yet, continue to fail.

Since writing this paragraph, I have made a concerted effort to photograph both Turkey Vultures and Black Vultures. As penance, I will include another picture of a Turkey Vulture below. Not a glamour shot, where the vulture looks its best, but rather a realistic picture of a vulture as it struggles through the trials and tribulations of life as all living things must.

Photo by Brian M. Wargo

This scraggly Turkey Vulture is having a rough molt.

Of course, I just lied! This is the ugliest vulture I have ever seen and wanted to share this picture because of the shock value, not because of the amazing wing structure that allows this bird to stay aloft despite having nearly half of its flight feathers missing. I wonder if vultures would ignore us or take advantage of us when we look our worst? Or will they just wait for our final demise and get their revenge as they devour our bodies.

This thought is revolting and reveals chasm that we have with nature—we only find parts of the natural world beautiful. I doubt that vultures worry about such esoteric ideas. But, maybe that is why vultures get no respect.

BROAD-WINGED SEASON

The middle of September is always special. The leaves are just beginning to change their color, the temperature is finally moderating from the scorching summer, and kids are back into the routine of school. But the real reason that September is special is because the Broad-winged Hawks are coming!

A surprising number of people go about their daily lives not knowing this fact. Schools do not proclaim this in their daily announcements, television newscasters seem oblivious to what is about to occur, and even the weatherman forgets to address this in his forecast. Instead, children hear about an upcoming assembly in support of the football team, television viewers will learn about the Great Pumpkin festival, and the weatherman will fill the elderly with fear about this year's snowfall predictions. Football is exciting, pumpkins are an interesting fruit, and snowfall helps recharge reservoirs, but really, who cares when you realize THE BROAD-WINGS ARE COMING!

It seems crazy that I would even have to explain what my excitement is about, but, for the uninformed, I will do my best. Chunky little hawks will be descending on your state like a flock of locusts. (Locust have a negative connotation, so let me try again.)

Small, pudgy hawks about the size of a crow will be traveling past your house without you even knowing it. (That sounds a little creepy or scary, so one more try.) A large number of moderate-sized raptors will silently and innocuously migrate past your state in an effort to find food for the winter months. (Wow, when I put it like that, it sounds about as exciting as the pumpkin festival.)

Photo by Jeanine Ging

A Broad-winged Hawk poses at Allegheny Front Hawk Watch.

What is it about the Broad-winged Hawks that makes them so exciting for hawkwatchers? The answer is multifaceted. The first reason has something to do with their organization. Broad-winged Hawks form large kettles, thereby amassing in large numbers. They find a thermal, congregate together, rise higher and higher until one of the hawks decides it is time to make a break for the next thermal, and then stream out one by one.

The Broad-wings Hawks make perfectly straight lines across the sky, either single file or making parallel lines with one another. The organization is military-like, with every hawk sweeping out a path that is motionless with respect to the bird in front or in back of it. The hawks do not change their posture, but remain completely frozen, almost standing at attention. This air show has the precision of a military march, minus the noise. The hawks silently move in unison across the sky, either at a constant or accelerating rate.

The flow of birds is relatively constant, which is a splendid change from earlier in the month and all of August. The Broad-wings force the hawkwatcher to constantly scan the sky, for a dozen or so birds could easily be missed. Making the same error in August would probably not have caused the count to be lowered because there were so few birds in the sky. Additionally, August birds are usually not serious about migration and rarely make such a deliberate line south. This is completely different in Broad-winged Hawk season, for the birds are on a mission and fly with intent.

Depending on the hawk site and the weather conditions, Broad-winged Hawks will appear to fly singly or bunch together. Often, the Broad-wings fly in squadrons, groups of hawks that collect and remain together for a few hours or even the day. The single birds that appear low and disorganized probably just ended their slow decent and are now looking for a new thermal. After one of the hawks begins to rise quickly, the rest swoop in to also get on that same thermal. They, again, rise and seem to be spit-out one by one.

Depending on how high the thermal lifted the hawks, they may have a little or a lot of altitude to lose before they need to repeat the process. If the birds glide downward for some time, they can often get farther and farther apart. To the hawkwatcher, this can give the appearance of a continuous line of single birds passing overhead.

This happened one day I was counting on the cliffs of Cumberland Gap Hawk Watch. The Broad-wings were kettling well before my position and they were not rekettling until they passed out of view over the next ridge. Therefore, all of the hawks were nearly single file directly overhead. To avoid getting the dreaded condition of "bird-neck," I decided to lay down flat on the cliff. I must have been lying there for forty minutes or so counting a fairly steady stream of reasonably spaced out single hawks when one of the immature Turkey Vultures landed about 10 feet from me on the edge of the cliff. I am not sure if he was just curious, checking to see if I was dead, or had just not noticed me laying on the edge of the cliff, but the young Vulture sat near me silently, looking me over. That was until I moved for the camera. At that point, he turned and jumped off the cliff, soaring away nonchalantly.

These Broad-winged days are so enjoyable because the birds are so ubiquitous that we do not look at individual birds, but just count them as numbers. And because we have such a large number of birds, we feel pressured to move on to the next bird as quickly as possible. As hawk counters, we get to click our clickers at a rapid-fire pace, spotting and clicking, spotting and clicking. We are in such demand, and we feel so necessary! For so long, we have been standing on the cliffs and mountainsides, staring at the barren sky, questioning our motives. But now, we are needed, we are important, and we are pushed to our limits. We have so many birds that hearing, "I got a bird out past the snag" is just as uneventful as hearing the words "I," "the," or "and"—just fillers in our sentences.

This is the real reason that Broad-winged Hawk season is so exciting, for it allows counters to feel the flow. It is about counting quickly and efficiently and moving on to the other birds. For a short time, we are immersed in purpose with no time for talking or exchanging stories, for these moments will be the stories in the future.

These are the times when hawk counting gives us purpose, direction, and conviction for what we do. Broad-winged season is the pinnacle of the otherwise lackluster hobby that others cannot understand. This is our time, our justification. Observers watch the hawkwatchers as they watch the Broad-wings pass and, for this brief moment, are actually envious. For this brief time, hawkwatchers are popular, in style, and important. We enjoy the show, because we actually become part of it for this fleeting moment.

THE OTHER WONDERS OF MIGRATION

The stability of each raptor in high wind is a clear indicator of its species. Eagles are stable, Sharp-shinned Hawks are not. Some wind is just too gusty, variable, and turbulent for comfort, even for eagles. On such windy days, I often worry that a bird will break its wing fighting the wind. Watching as the birds get thrown around is unsettling for me and I am on the ground. It is on such days that I am always amazed to see the most unlikely and unstable of all the migrators, the Monarch Butterfly. Somehow, these featherlight flying creatures are able to survive up in the air even during extreme wind. They flap their oversized wings in an effort to …actually, I do not know why they are flapping, for the wind seems to dictate where they are going. I have often wondered what would happen if they stopped flapping.

On windy days, leaves often get pulled off trees or directly from the ground and become airborne. Occasionally, leaves will be pushed up the 200-foot cliff at Cumberland Gap and fly straight over my head, gaining altitude until they get out of sight. Are Monarchs similar to leaves in the wind? They can't be. Monarchs migrate and leaves don't. Somehow, these insects that weigh about the same as an eagle's feather power themselves towards their destination. Even more miraculous is that none of the butterflies that begin the migration actually make the round-trip. The children of the children of the children complete the full trip.

However they do it, Monarchs can be as exciting as the larger hawks when they migrate. One late September day, I only counted 16 raptors. It was a nice variety that included a Northern Harrier, a Cooper's Hawk, Red-tailed Hawks, Sharp-shinned Hawks, and Broad-winged Hawks. But the day was really special because 193 Monarchs passed me on their way south. Every few minutes, a new butterfly would appear out of the ether and pass me. I continuously attempted to capture their flight with my camera, having it on and ready, but the Monarchs would appear and then disappear before I could focus on them.

Photo by Brian M. Wargo

Similar to the hawks, Monarch Butterflies migrate north and south.

Even on this rather calm windy day, the Monarchs were moving surprisingly fast. Each one took a different path, but they all moved directly south. Some were in better shape than others. I remember one that was missing part of its wing and was struggling a bit, but it, too, kept moving. Every one of the bright orange and black insects were stunning and seemed to make the day brighter and cheerier. By the end of the day, I was looking forward to a record number day.

When the Monarchs finally stopped moving for the day, I was ready to call it a day myself. That is when I began tallying the day's totals. It was then that I looked at the Blue Jay clicker and recognized that they also had a record day with 139 Blue Jays in total. I remember watching them as they passed. They would move in small groups of seven or eight and would collect on a pine tree. Their iridescent blue strongly contrasted with the green needles, giving the appearance of a beautifully decorated Christmas tree.

When a requisite number of the Blue Jays accumulated, they would begin to fly off across the gap at Cumberland. Their quick wing flaps seemed ineffective at propelling them to high speed. Instead, they slowly moved across to the other mountainside. Unlike other birds that seem to fly quickly and effortlessly, the Blue Jays seem to work hard the entire time they are airborne. Regardless, they certainly do migrate far distances and in huge numbers. Although 139 birds is a low number for many sites, seeing this many Blue Jays at Cumberland Gap was exciting.

This may be the ultimate lesson that hawkwatching teaches us—to be aware of the wonders that go on around us all the time. By cueing our senses to the workings of the natural world, we begin to find the extraordinary in the ordinary. And when the ordinary changes, we are primed to find even more excitement.

This was the case when I arrived at the Cumberland Gap site one morning and discovered that the main road under the hawkwatch had a backed-up storm sewer that flooded the road. In order to fix the problem, the road was shut down for a few hours. The difference that this made with the ambient road noise removed was marvelous. I could hear all kinds of birds from far away that would normally be drowned out by passing cars. For about two hours, the site was free from motor vehicles (trains, trucks, cars, motorcycles, etc.). The noise level is never loud on a typical day, but the attenuation of extraneous sounds made for a pleasant morning, despite the falling drizzle and the corresponding low number of birds.

I now notice and enjoy the smaller aspects of my surroundings and find joy in watching nature as it unfolds. Dropping a crumb from my sandwich or stepping on a stinkbug initiates a set of events that I had been oblivious to previously. Scout ants find the bread and dead insect within minutes and begin in earnest to move their treasure. No matter what organic material hits the rocks, the lone scout ant will begin to clean it away. I always intend to discover where they take their spoils, but I need to scan the sky and, by the time I remember to look for the ant, it is too late. They are gone without a trace.

Photo by Brian M. Wargo

A Fence Lizard watches for hawks on the cliffs at Cumberland, MD.

I now also recognize that I am not the only hawkwatcher on the cliffs. A multitude of other creatures also scan the skies for hawks. One of them is a Fence Lizard that would appear out of nowhere and sit next to me. On most warm days, I expected the lizard to make a showing, although he also arrived on some surprisingly chilly days. He would tilt his head and point one of his eyes to the sky anytime a bird would pass overhead. Occasionally, he would get excited about a bird, but would normally ignore them unless they came too close.

After many days, the lizard began to feel like an outdoor pet. I pushed my luck one day and petted him on his back. He clearly did not like us taking the relationship to this level, but he recovered pretty well from the incident. It also seemed like every time I would eat, he would begin to eat, much like pets with their owner. Insects would be in front of the lizard's face for an hour, but it was only when I began to eat that he showed any interest in his meal. Surely, it was because he finally warmed up enough to chase the prey. Although, I wonder!

Ultimately, hawkwatching is an activity that forces one to stop and connect with nature. The hawks may draw us in, but the rest of nature keeps us coming back. The multitude of spiders, snakes, clouds, leaves, worms, lizards, moths, rocks, porcupines, etc. ensure that there is always something worthy of our attention. We just have to open our minds to the dynamic and wondrous world we co-occupy with so many interesting living and non-living entities.

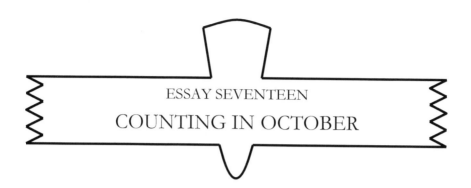

ESSAY SEVENTEEN

COUNTING IN OCTOBER

It was a homecoming of sorts. I was returning to Allegheny Front Hawk Watch seven years after my first visit. This is where I had started hawkwatching. I know this because my son Theo's first big outing was to the Allegheny Front, and he was only 12 days old. That day marked a significant shift in my birding. Like so many other birders, big raptors were not foremost on my mind. Yes, raptors were birds, but those giant killers were rare and, when they showed themselves, they threatened the small, delicate, multicolored birds that captured my attention.

As a beginning birder, I focusing my binoculars on the birdfeeder that I religiously filled each morning. Accipiters, such as Sharp-shinned Hawks and Cooper's Hawks, terrorized the feeding birds. More than once, I witnessed their deadly attacks. The victims had been the stately Northern Cardinal and the dainty American Goldfinch, and I always feared for the much too trusting Black-capped Chickadee. The mere thought of something killing my beautiful feeder friends was reprehensible, especially coming from a traitorous fellow bird.

When an incident occurred, I felt guilty and obligated to formulate a plan so as to ensure such an act would never happen again. Safety measures, such as relocating the bird feeder next to thick brush, thwarted the birds of prey. Feeder birds simply jumped to the safety of the tangled thickets, thereby nullifying the danger.

I enjoyed seeing raptors, but certainly not at my feeder. I did not want to be complicit in a murder of the innocent. If a death occurred, it was on me because I had lured birds in with delicacies that they simply could not resist. Unshelled sunflower seeds, peanut butter, lard, and thistle offered in gluttonous quantities, in a lavish pavilion with a perfect perch, all under a protective roof. To the birds, I was the giver of life—I was a god!

I selflessly offered sustenance, love, and compassion. The birds recognized me, learned my schedule, and waited for me near the feeder. They rejoiced as I approached, knowing that I carried the calorie-rich delicacies they sought.

Photo by an Allegheny Front Hawk Watch Member

Our first trip to Allegheny Front Hawk Watch was in 2009.
From left: Jeanine Ging, Theo, Meadow, and the author of this book

The birds of prey also waited, perching in the darkness of the trees, waiting for a careless bird. These birds of prey were ruining my status as a deity. Conspiratorial feeder birds might begin to think of me as a wolf in sheep's clothing! I was peddling gluttony in exchange for their lives. The raptors turned my solemn act of kindness into a trap that would kill.

This was my mindset as a birder. I had forgotten the birds were not here for my enjoyment. They were struggling through life as all other creatures must. Their beauty, fitness, and habits were not for me, but for them. They were the product of nearly infinite choices that nature had selected generation after generation. It just so happened that I was able to enjoy the spoils of this evolution in the form of perfectly adapted flying creatures.

My first trip to the Allegheny Front Hawk Watch showed the raptors in a new light and forced me to revise my convictions. I felt young again, not in a youthful exuberance fashion, but rather in a nativity-meeting-reality manner. This happened before. As a kid hearing about the Vietnam War, I was disgusted hearing about the destructive force of a bomber or a jet fighter. The thought of bombing or killing other human beings on a large scale was vile and undignified. Later, as a young man, I actually saw military craft up close. It was captivating, wondrous, and, dare I say, beautiful. This caused cognitive dissonance, for I simultaneously felt awe and unease. The perfect design juxtaposed with the intent of the craft was paradoxical. Perfectly destructive was an unusual set of descriptors that I had trouble situating in my overall psyche.

As a contemplative being, I was torn: Was I becoming a man by accepting the duality of destruction and perfection or simply turning my back on the notion of good and evil? The latter narrative was cleaner and easier to categorize while the former was complex and fuzzy. Passing from childhood to adulthood forces just such conundrums to the forefront of our moral thinking, and my visit to the Allegheny Front was just such an event. I had learned over the years that we must let go of our absolutism and accept the dichotomy of our reality. The trick is to somehow keep our humanity without becoming nihilistic or apathetic!

I was a grown man at this point and needed to come to grips with birding. I was not actively engaging with nature as I had felt and needed to accept that I was simply a voyeur, watching nature play out her game. This was clear after viewing a Merlin bounce by as I stood on the plateau of the Allegheny Front, for I clearly understood what it must be like to be ambushed. The bird seemed to come out of thin air, pump a few athletic wing flaps, and disappear. If I was a songbird, I would have no chance against such a perfect fighter.

Similarly, I recall seeing my first Golden Eagle. It soared passed the hawk site without flapping its wings. The bird stoically looked down upon me as it drifted across the sky. From my vantage point, the darkened silhouette was like a bomber—appearing slow and steady, but with freighting potential. This eagle must have sent shivers down the spine of other birds and small mammals, just as a bomber would do to civilians when it flew overhead.

Every raptor that I saw was suddenly impressive, fitted with their unique style and killing ability. I gained respect for the raptors because of their flying capability, great speed, and design. All of these attributes necessarily meant the death of another. How odd it was that I was a certified biology teacher and had kept my two great loves artificially separated: science and nature. Clearly, I needed a reckoning between my ideals, my hobby, and my academics.

All of that was in the past. I was returning to Allegheny Front Hawk Watch not as a naïve birder, but as a counter. Clearly, I overcame my trepidation concerning birds of prey. I had spent three seasons at Cumberland Gap Hawk Watch counting and intently studying raptor identification. The hundreds of hours spent there had markedly improved my hawk skills. I was now able to not only identify species, but age and sex them reliably. I was confident in my ability as a hawkwatcher and as a hawk counter, but felt like I was now working in the big league.

The move from Cumberland Gap to Allegheny Front was spurred by recent health issues with a few of the counters at Allegheny Front. The other counters who were filling in were feeling the extra load and thought they could use an additional counter. At the same time, Cumberland Gap was becoming less and less safe. I finally decided that the risk of going to Cumberland Gap was just too great to justify the data that I was collecting. The timing was right and the transition was quick. I would take over as the Saturday counter.

The Saturday count is, depending on your perspective, the most cherished or the most dreaded day to count at a hawkwatch. For many, Saturday is a day off and the only day that they can watch hawks. This means that people will be around on Saturday. Sometimes, lots and lots of people. This is distracting to some and a godsend to others. Weekday counters are either alone or have a few of the regulars that stop by. Saturday is a time to socialize with others as they practice their craft. Regardless, Saturday is different than all the other days of hawk counting. And, if the weather is conducive for a large flight, Saturday can draw an impressive number of respected birders.

My first official Saturday was just such a day. By eight in the morning, serious hawkwatchers were unloading their vehicles. The winds were out of the east and strong. It was the end of October, and the peak of the Golden Eagle season was upon us. The week prior, 74 Golden Eagles had passed in a single day, shattering the east coast record of 61, which was also held by the Allegheny Front. This week looked like it might be another impressive day. By 10 a.m., the field was filled with antsy hawkwatchers. Everyone was primed and ready for the flight. Spotting scopes perched, binoculars in hand, and cameras with massive 600 mm lenses mounted on Gimbal heads.

I was a familiar face at Allegheny Front, but not as a counter. Seven years ago, I could not positively identify a Turkey Vulture, let alone age a Golden Eagle. Now, I was to make the official call and enter my judgement into the record. I was working in front of many of the hawkwatchers that were already at an expert level when I first came to the Allegheny Front. Many of the old-time counters from the past showed up because of the favorable conditions. I must admit, I felt a lot of weight on my shoulders. There were at least 25 sets of experienced, high-level hawkwatchers at the site. This included counters from other sites who had heard about last week's big day and wanted to experience what Allegheny Front had to offer.

The flight started early and the birds began coming in. One of the first birds of the day was an accipiter. It was not a large bird, so it most likely was a Sharp-shinned Hawk. Everyone raised their binoculars and focused on the bird. It was low and moving fast directly overhead. I heard a few mumbles of "It looks like it might be a Cooper's Hawk." Cooper's Hawks are sexier than Sharp-shinned Hawks—they are usually a bit larger and they're rarer. It is always more exciting to write that you saw a Cooper's Hawk than the more ubiquitous Sharpie! This accipiter appeared large; however, it was flying alone, so it is problematic to know if the bird was large. It was not flapping and was diving down in a bit of a stoop, thereby negating the technique of looking for a shrugged look or a larger-headed profile.

Photo by Brian M. Wargo

Sharp-shinned Hawks look pudgy compared to Cooper's Hawks.

Regardless, this larger bird appeared to have the tip of its tail cut with a sharp pair of scissors that had that had just been pulled out of a black paint container. This sharp tail with hardly any white at the tip was diagnostic that it was a Sharpie.

One of the veterans yelled, "What did you mark that as?"

I replied, "She was a Sharpie!"

He yelled back, "Good, cause it was a Sharpie!" He was then about to say something, which I bet was a statement concerning the larger size of a female Sharp-shinned Hawk versus the male, but then recognized that I had already preemptively put that in my statement. He closed his mouth, deciding not to say anything else, smiled and nodded with approval.

The next hour, a Red-tailed Hawk came through and it was quietly marked. Then, an eagle appeared right off the edge of the ridge and quickly made its way around the southern edge of the site. My heart raced. I barely got a glimpse of this bird. It had some white in the tail and what appeared to be whitish on its back. He must have just taken flight and was only about 70 feet off the ground, which was just below eye level of the hawkwatchers. As it was going out of view, I clicked off a picture so as to analyze it.

The picture was not very good. The whitish on the back was not very white and the tail was difficult to see. My heart sank! I had just seen an eagle right in front of me and I could not distinguish if it was a Bald or a Golden. Everyone looked to see what I would say! I wanted to say, *"I am a fraud! I can't identify raptors. You should find someone else to take my place."*

Instead, I quickly asked, "Who had a good look?"

To my amazement, I heard, "It may have been a Bald," and from another, "Actually I thought it was a Golden." Clearly, others had been equally confounded by this bird.

I ran over to the guys with cameras and asked if they could pull up the pictures and they gladly did. One camera caught a good shot of the tail, another of the axillaries just as the bird got to the top of the wing stroke as it flapped, and yet another captured the top side of the bird as it stoked downward. It turns out that the whitish on the back was actually tawny in color. The axillaries did not show any white and the tail had more of a white wash. It was a sub-adult Golden Eagle. Everyone felt a sense of relief, but none more than me. I almost had to put down the dreaded UE – Unidentified Eagle. I was so grateful to have the cameras there. I thanked each cameraman and recorded the bird as sub-adult Golden Eagle.

Everyone talked about that eagle for the next 40 minutes or so, going over the fine points of eagle identification. Several members came up to me and offered a tip and pat on the back, agreeing how perplexing that bird was despite it being so close. This discussion stopped promptly as an adult Golden Eagle flew by at 10:05 a.m. followed by a juvenile Golden Eagle at 10:08 a.m. Three more adults came through at 10:18, 10:26, and 10:59 a.m. In addition to these eagles, a Merlin, six Red-tailed Hawks, a Red-shouldered Hawk, a Turkey Vulture, and four Sharpies also came through. The next hour, another Golden Eagle, 44 Red-tailed Hawks, five Sharp-shinned Hawks, five Turkey Vultures, and an American Kestrel passed.

At noon, the Golden Eagles began to pick up again, with an adult at 12:00 p.m., followed six minutes later by a juvenile. Three minutes later, another adult appeared. This very rapid set of Goldens is challenging because each eagle is marked on separate eagle page that requires additional information, including the age of the bird, the height of flight, and the direction of flight. Keeping each bird straight, marking it down accurately, and counting all the other raptors that are also migrating can be daunting.

This was certainly the case during this hour because, at 12:33, another eagle came through. This particular bird had white covering his abdomen and some in his axillaries, clearly indicating he was a sub-adult Bald Eagle. This eagle had just been processed when, at 12:43, another adult Golden Eagle came through, followed by another sub-adult about 10 minutes later. Immediately after identifying this bird, another appeared right behind him. It was an adult. He was marked as coming through one minute later at 12:56. The hour was just about up and it was almost time to get a weather reading. During this hour dominated by eagles, 34 Red-tailed Hawks had also passed through as well as another four Sharp-shinned Hawks and a Turkey Vulture.

As I was writing the hourly weather summary while sitting at the picnic table, someone yelled, "Here comes another one." I scrambled to my feet, grabbed my binoculars, focused and... It was behind the tree line. I lost the bird.

I quickly yelled, "Anyone get that bird?"

Thankfully, three cameras had taken nice shots and, using only one, it was quickly assessed to be a sub-adult. I thanked the camera operator and began heading back to write down this eagle's information when another eagle came overhead. This one I got a good look at and it was a sub-adult Bald Eagle. I grabbed my phone to check the time and hurried back towards the picnic table, almost breaking into a run. It was almost comical that I was trying to keep up with all the eagle activity, but was not even done with hourly weather data.

Photo by Brian M. Wargo

Not all Golden Eagles are as accommodating as this bird.

Everyone was enjoying the wonderful stream of the giant raptors. Smiles were worn by everyone and the picture of hawkwatchers laughing and putting their hands on each other's shoulders is one that is burned into my brain. The atmosphere was festive with everyone having a great time. I, however, was running like a chicken with its head cut off, racing to get the information onto the data sheets before forgetting about it. So many eagles and other birds were flying over that some birds passed without anyone verbally acknowledging them. The last eagle was one of those birds.

Ed Gowarty, a veteran at the hawkwatch, yelled to me, "Did you mark that second Golden Eagle?"

I didn't know he was kidding at the moment, and I stopped and said, "No! That was a sub-adult II Bald Eagle."

He replied, "Are you sure about that?"

I said "Yes!" He and several of the other old-timers began to laugh as they pointed at me. He, of course, was just kidding. He knew how chaotic the situation was and was testing to see if I was handling the pressure. Tom Kuehl, who has counted for Allegheny Front and is the former president of the Pennsylvania Ornithological Society, chimed in between the laughter and gave a sincere gesture of gratitude.

"Brian, thank you for doing this! I am sure glad I am not in your shoes today!" With that, he grinned and looked back at the sky and yelled, "Sharpie!"

As I recorded the two eagles at 1:02 p.m. and 1:04 p.m., I recognized that I had not yet eaten lunch. Jeanine, my significant other, had put food on the table without me seeing it, so it was perfect timing. I quickly and nervously ate, scanning the sky the whole time. As luck had it, the eagles took a half-hour break. Red-tails and Sharpies were coming through, though. This was not a big deal, for I had a clicker in each hand and could continue eating as I clicked each bird that passed. But just as I put the last bit of food into my mouth, I noticed that one of the Red-tailed Hawks looked odd. His flapping was Buteo-like, but something was different about it. I locked onto it with the binoculars, but it was stooping, so very little was offered for identification. Suddenly, the raptor spread its wings and tilted slightly. The top shoulders of the bird were a beautiful reddish-brown hue.

That's it, I said to myself. It was a Red-shouldered Hawk. Out loud, I yelped, "Red-shouldered." Everyone looked towards this inconspicuous bird, but it had gone back into a stoop at that point, so no one was able to get the view I got.

Someone yelled, "That was a Red-shouldered?"

"Indeed!" I replied.

Immediately after the Red-shouldered appeared, an adult Bald Eagle floated by. This easily-identifiable eagle slowly moved across the sky and seemed to slow the pace of the count. It was a welcome relief. For the next half hour, the count drew down to a trickle. At the 2:00 hour, I was able to get the weather summary without a rush and was even able to chat with many of the observers. Another easy eagle ID at 2:09 with a Juvenile Golden Eagle began an hour-long hiatus for the eagles. Other birds were coming through, including 33 Red-tails and two accipiters. The first was a Sharp-shinned and the other a Cooper's Hawk. Two small falcons came by and I missed one completely. Thankfully, others had a good look and the consensus was an American Kestrel. I saw the other one, and that rounded out the hour.

The 3:00 p.m. hour showed a decrease in the Red-tailed flow, but the Goldens kept coming: a sub-adult at 3:07 p.m. and then a juvenile at 3:29 p.m., followed by another sub-adult at 3:34 p.m. The four o'clock hour showed that the pipe was running dry. Then, at 4:24 p.m., an adult Golden Eagle came directly overhead and low. Another adult came overhead at 4:59 p.m. This seemed like a nice end to the day, for 5:00 p.m. is the standard time the hawkwatch stops for the day.

Everyone began taking their belonging to their cars, and some people had already departed. But then, two minutes after five o'clock, a juvenile Golden Eagle reared his head over the tree line. Suddenly, hawkwatchers that had to leave decided to stay. I, like everyone, assumed that the flight would be over by this point, but I was going to stay until at least 6:00 p.m. The eagles did not disappoint. At 5:24 p.m., another sub-adult Bald Eagle passed overhead. Finally, at 5:47 p.m., the last eagle of the day passed north.

As everyone was getting ready to leave, two more Golden Eagles were coming over the trees. In an excited state, one of the senior members ran over towards his spotting scope. Because it was late in the day, the eagles were very low and obviously looking to set down for the night. The sight of a man running in their direction caused both birds to quickly veer off course, making a U-turn. The rule at Allegheny Front Hawk Watch is that only birds that pass from the north to the south are counted. Therefore, these two birds were not counted.

Despite the unfortunate ending, the day was a huge success. Everyone who showed up witnessed a wonderful flight, with a total of 23 Golden Eagles, four Bald Eagles, 162 Red-Tailed Hawks, 23 Sharp-shinned Hawks, three Red-shouldered Hawks, one Merlin, three American Kestrels, a Cooper's Hawk, and eight Turkey Vultures. Almost everyone who stayed made a concerted effort to shake my hand and express their approval.

As the last cars drove away, I felt an odd sense of warm joy coupled with exhaustion, which mitigated the cold that had chipped away at my body all day. That night, as I sat in the darkness at the picnic table trying to get reception to enter the data into hawkcount.org, I reflected on the full circle that I had just completed. I had returned to the place I had started hawkwatching, no longer a novice, but a legitimate member of the core team that ran the show. I was now the hawkwatcher that I had dreamt of becoming. It was a perfect homecoming that I will cherish forever.

ESSAY EIGHTEEN
SOARING: IT'S NOT MAGICAL –
IT'S JUST NATURAL

A man and Red-tailed Hawk are standing in the middle of the plains of Kansas. The world is seemingly flat and appears to extend forever. The wind is howling, and the man decides he has had enough. Nothing is happening here, and he is starving. He begins to walk into the wind in an effort to be somewhere else.

After walking some time, he looks back only to see the Red-tailed Hawk remaining at the same spot. He walks farther, now noticing some small animal ahead of him. He is so hungry that he welcomes the thought of eating a ground squirrel or a rabbit. He knows that he is just not fast enough to catch the animal. He sits, trying to form a plan.

As he thinks, he looks at the Red-tailed Hawk, who suddenly extends his wings. Almost magically, the hawk ascends nearly straight up. It then gets higher and higher. After reaching a significant altitude, the Red-tailed Hawk dives, gains great speed and makes a straight line for the ground squirrel. In a matter of seconds, the hawk has a meal.

The man becomes contemplative, thinking through the situation, feeling that he has witnessed a miraculous event. His large brain allows him to think about his thinking and his existence. He feels both wonder and melancholy. He wishes that he, too, was able to levitate from the earth. He is but a man, he realizes, lacking the superpower of flight. He then moves on towards the sunset, carrying his enormous brain and his empty stomach.

So, how does it happen? How does a bird magically lift upwards when wind travels across the wing? It's not magic, just physics. To understand this is to understand what constitutes air. All matter, including air, is made up of atoms—most likely pairs of atoms. Nitrogen, for example, pairs with other nitrogen to form N_2 and constitutes about 78% of all the molecules in the atmosphere. Oxygen also forms a diatomic molecule (O_2) and accounts for 21% of the atmosphere. Carbon in the form of carbon dioxide (CO_2), hydrogen, argon, helium, and other elements are found in trace amounts.

Regardless of their chemical differences, gas molecules act similarly as physical objects in air. They move around at staggeringly fast speeds. Oxygen molecules, for example, are traveling at around 1000 mph. Nitrogen is even faster. We do not notice this because they are so small and there are so many of them. Just to put this into perspective, there are about a octillion—that's 1,000,000,000,000,000,000,000,000,000 (10^{27})—molecules of air in a standard-sized room. Each of these molecules travels only a short distance before bumping into another molecule, thereby changing its direction. Every one of these collisions has momentum that can cause the objects they collide with to move. The number of collisions that hit a particular surface is called pressure. If there are more molecules hitting a surface at any given time, we say that more pressure exists.

An air-filled balloon keeps it shape because of the gargantuan number of molecules hitting the insides of the balloon at any time. We do not see any individual hit because the molecules are so small. The balloon looks smooth and static, but at a molecular level, it is a chaotic bombardment of tiny particles smashing into one another on the inside of the balloon. There are also molecules hitting the outside of the balloon, just like the inside. Latex balloons are composed of an elastic membrane that squeezes the molecules together a little bit tighter than the molecules outside the balloon. For this reason, the molecules inside are slightly more concentrated. The balloon, therefore, is denser than the surrounding air and falls to the ground.

A bird extending its wing has quadrillions of molecules hitting the underside of the wing. It also has an equal number hitting the top side of its wing, thereby having no net effect. But when air travels past the wing, different numbers of molecules hit the top and bottom of the wings. This is because a wing is not symmetrical. It tends to be thicker at the front and thinner at the rear. It is also reasonably flat on the bottom while bulging on top. As air reaches the front of the wing, it will either travel along the curved top portion of the wing or the straight bottom portion. Straight lines are the shortest path between two points, so the air molecules that travel along the underside of the wing are traveling a shorter distant than those above.

The air molecules above the wing are trying to reach the end of the wing at the same time as the molecules at the bottom side. This means that they have to travel faster to arrive at the end of the wing. It also means that there will be a larger spacing between each of the molecules. This extra space means fewer molecules per area of wing; therefore, there will be fewer hits on the top of the wing compared to the bottom of the wing. In summary, it is the shape of the wing that forces the molecules of air to move along the top side of the wing in a rarified form that creates lift.

The amount of lift is dependent on the speed of the wind, but also the composition of the wind. The relative components of air remain pretty much the same, but each day, the amount of water vapor in the air does change. Hawks do not seem to like soaring in humid weather. The reason is also related to the molecules that make up air.

In any give volume of air at a given pressure and temperature, there are a certain number of molecules, regardless of their type. To be technical, there will be 6.02 x 10^{23} gas molecules in 22.4 liters at standard temperature and pressure. Oxygen is more massive than nitrogen, which is more massive than hydrogen. The periodic table gives the values for the mass of each element, and it is easy to get the molecular masses. For example, a single nitrogen atom has a relative mass of 7 g/mole. Diatomic nitrogen (N_2) is a molecule of two single nitrogen atoms and, therefore, has a mass of 14 g/mole. Despite N_2 being made up of two atoms, it functions as one particle in a gas. A water molecule is two hydrogen atoms combined with one oxygen atom. Oxygen has a mass of 8 g/mole, and hydrogen just 1 g/mole, giving a total molecular mass of 10 g/mole. Diatomic oxygen (O_2) has a molecular mass of 16 g/mole.

Photo by Brian M. Wargo

The feathers on this immature Turkey Vulture are bent upwards from the difference in pressure above and below the wing.

As the number of water molecules increases, less of the air is composed of the more massive N_2 and O_2 types. This means that humid air is less dense than dry air. This is counterintuitive, for when it is humid, humans feel that the air must be more dense, when in reality, it is less dense. Less density means less powerful hits of the molecules on the wings, and, therefore, less lift. This difference in lift may be small, but to an expert flyer, it is noticeable.

Similarly, birds also will have variation of lift with atmospheric pressure. It may seem that pressure differences would affect the bottom and the top of the wings equally. That is true, but it is the difference in pressures that matter here. Thinking of scale may help. For example a nine-to-ten ratio can be written as 9/10, 90/100, or 900/100. But, the difference between 1 and 10 is 9, while the difference of 10 and 100 is 90. So, if the wing shape gives us a ratification factor of 9/10, the total lift will depend on the pressure, which, for a gas, corresponds to the number of molecules hitting the wing.

Raptors do not understand any of this. They can't read, and they lack any type of semiotic system of communication. They may even be considered less bright than other birds. Humans are advanced in their cognitive abilities and can prove these principles scientifically. But, in the end, we are forced to dream about flying like a hawk. Maybe the hawk dreams of understanding science. Most likely, we are overthinking this one.

Kevin Georg may have counted more hawks than any other person in 2015. The total number at his two primary posts, Mackinaw Straights in Michigan and Corpus Christi in Texas, gives a value greater than 700,000 hawks for the year. Over 500,000 of these birds were Broad-winged Hawks, and in just a two-day period, he counted over 148,000 Turkey Vultures, ending the year with approximately 175,000 of these vultures. It is hard to fathom counting almost three quarters of a million of anything, let alone hawks.

But don't be fooled; Kevin has not just seen the big-number hawks, he has also seen a pair of Harris Hawks, a pair of Short-tailed Hawks, four Crested Caracara, five White-tailed Kites, 13 Zone-tailed Hawks, four Northern Goshawks, 43 White-tailed Hawks, 89 Swallow-tailed Hawks, four Ferruginous Hawks, and four Prairie Falcons. In total, Kevin has counted over 27 species of raptors. So, how does he do it?

Geography is a key player in Kevin's huge numbers. Both Mackinaw Straights and Corpus Christi are natural funnels for migrating hawks. Corpus Christi is near the Gulf of Mexico, just west of the Aransas National Wildlife Refuge. Hawks from the Eastern half of the United States fly south, hit the Gulf and follow it around into Mexico. Soaring birds are seemingly attracted to this water/land divide, but the reality of situation is that they do not get thermals flying over the water. They, therefore, stay near the land. The hawkwatch sits on the highest parcel of land in the surrounding area and gains a few more feet with an elevated viewing deck. The hawks are almost forced to travel directly over top of this site.

Corpus Christi is a hawkwatcher's heaven...and hell. The bird numbers are incredibly high, but so are the temperatures. On an average September day, temperatures can rise from an overnight low of 72 degrees up to 91 degrees by afternoon. On hot days, the temperature may be higher than 105 degrees. If you are thinking that this temperature will feel nice because the humidity is low, think again. The average humidity is 75%.

The confluence of high temperatures and high humidity coupled with the lack of natural shade equates to birding in a sauna. When I ask Kevin how he survives, he replies, "I eat crackers."

I excitedly respond because I know what he is getting at. I reply, "Cause it makes you drink more!"

"That's right. All day long, you have to drink," Kevin responds.

I ask the obvious question, "What about all the bathroom breaks?"

He laughs, "You sweat it out, so it's no sweat." We both laugh at the play on words.

At the other end of the United States sits another extreme weather area: the Mackinaw Straights in Mackinaw City, Michigan. When the count starts in March, the average high temperature for the day is below freezing. The low temperatures hover around 20 degrees and can become downright dangerous when a cold snap hits. When combining the high temperatures of Texas and the low temperatures of Michigan, Kevin may have a 130-degree temperature swing. The real feel temperature difference may be more like 170 degrees.

The geography of the Great Lakes is a giant cone for the soaring hawks as they travel north. Birds traveling through Indiana and Ohio get trapped between Lake Huron and Lake Michigan. The northern tip of land is where the hawkwatch sits. It is here that the birds will cross, for the water narrows down to five miles. Both birds and humans can see the land on the other side. Humans can take the Mackinaw suspension bridge, which happens to be the longest of its kind in the Western Hemisphere. Some people get spooked and will not cross the bridge once they arrive. The same happens with the birds. They fly out over the water only to turn around. The Mackinaw Straights is both a thing of beauty and a thing to fear.

Kevin Georg is like the birds he counts. Each year, he migrates from the north to the south of the United States and then back north again. He endures similar weather as the birds and seems to be in tune with the natural flow of things. Just as many ask how the birds complete their migration, other hawkwatchers often wonder how Kevin does it.

Kevin Georg is a tall guy with long hair, who at first looks imposing, but then speaks with a surprisingly mild and comforting voice. He may best be described as a hippie who stumbled into being the man who watches the most hawks. Kevin's story began 30 years ago when he took his Husky dog to a veterinarian. The vet and Kevin spoke about their shared love of the outdoors. With that first engagement, Kevin and the vet began watching hawks on the various ridges around Pennsylvania.

Like so many hawkwatchers, Kevin and the veterinarian were hooked right away and began spending more and more time looking at hawks. As the hawkwatching bug developed, more and more time was spent thinking about how to allocate more time for watching hawks. The factors that impeded nonstop hawkwatching were typical life issues and norms, such as maintaining a job, raising a family, having a bed to sleep in, having a mailing address, and having money to eat.

Hippies always seem to find a way, and Kevin would find a way to watch hawks. Describing Kevin as a hippie is not entirely accurate. Kevin was never really a hippie, just hippie-like. Unlike the negative stereotype, Kevin worked hard in the steel mill and enjoyed the challenges he faced each day. As a millwright, he would climb tall towers hovering above the inferno of molten steel, operate equipment that could kill a man in a second, and ran massive cranes.

Kevin enjoyed these challenges and marveled at the processes he was exposed to at the mill. His real dream was to become an astrophysicist like Carl Sagan, but those dreams were dashed when his father got sick and he became the man of the family.

Living in Johnstown meant jobs in the steel industry, and that is where Kevin went to make money for the family. That all came crashing down one day, literally, when the crane he was operating fought back. As Kevin was trying to open the door of the cab, the door gave way, and he and the door crashed down below, breaking his back. Thankfully, it was not thought to be too severe. Before long, Kevin was back at work. However, something was wrong. The pain never subsided, and it would rush down to his legs without warning. It soon became clear that he had nerve damage. He persevered under light duty for some time until a few more minor accidents revealed that Kevin was disabled, and his back injury was inhibiting him from being able to carry out the work safely. They asked him to take disability, which he did. Unfortunately, the pain never took a hiatus.

After receiving good marks on a battery of aptitude tests, Kevin was authorized to attend college. It was a real consideration, except for the fact that reading only exacerbated the pain. Instead of taking his mind off it, it just sat next to him, page after page. College was not going to work out. As Kevin was soul-searching, which usually involved watching birds, he serendipitously realized that his brain was distracted every time he looked through binoculars. For whatever reason, this trick worked. If the pain got worse, Kevin would look through his binoculars more intently, almost as if transmogrifying the pain out of his body and into the sky. It was weird, but effective! Kevin found that getting into the zone of the sky distracted him from the pain, and in the process, created a sky-scanning machine.

For those not used to looking through binoculars, it is an odd experience. Your brain finds it strange to look at the world through a different magnification. After some time, your brain informs the body that it requires adjusting back to default conditions, that of normal vision. If unheeded, the brain begins punitive actions including inducing pain in the head, eyes, and/or the neck. It is best to heed this warning or risk destabilizing the body in a woozy vertigo.

This is the main reason that people look through binoculars for a few seconds and then put them back down. Occasionally, through genetics, biofeedback, or sheer will, certain individuals overcome these impediments and can continuously look through binoculars. Kevin Georg is one of those individuals. He can deep-scan the sky without break for 20 minutes or longer. This attribute allows Kevin to see birds that others have little chance of detecting.

Kevin Georg is a hawkwatcher, but his specialty is scanning. For those not into hawkwatching, a scanner is a hawkwatcher that searchers for birds by using magnification devices to find what the naked eye cannot readily see. All hawkwatchers have a propensity towards a particular aspect of hawkwatching—Kevin's is scanning.

Other hawkwatchers are excellent spotters, meaning that they look at the sky and are able to notice slight changes in the background. This is natural for some, but not for all. A good spotter looks around, notices that a small speck they saw just a moment earlier has moved. They then follow the speck, and it usually turns into a bird. Other hawkwatchers are excellent with their identification skills. Sub-groups include plumage specialists, shape specialists, and behavior specialists. All hawkwatchers need to be proficient with all of these aspects, but again, everyone seems to be better with some aspects over others. Factors that influence these attributes include one's ability to distinguish colors, attention span, being near or far-sighted, being a visual learner, distractibility, and overall understanding of birds, hawks, and weather phenomena.

The merger of receiving therapeutic effects scanning the sky, combined with Kevin's love of watching birds and his newly found free time, set in motion a series of events that would not only make Kevin one of the best scanners in hawkwatching, but would also land him a spot at one of the highest count hawkwatches in the country.

Corpus Christi is a fitting place for Kevin and his skill set. He can sit for marathon sessions, deep-scanning the sky, counting huge numbers of birds. I personally admire this attribute, probably because it is where I am weakest. My strength is in identifying individual birds once they are spotted. In between bird identifications, I unfortunately require frequent breaks to stave off light-headedness. Just the thought of Kevin staring through his binoculars for hours at a time makes me a bit dizzy.

Rarely does a hawkwatcher have specialty in all areas, but it does happen. My mentor, Tim Anderson, was one of these odd individuals. He could spot, scan, and identify effortlessly. He was good with the naked eye, binoculars, and a spotting scope. Unfortunately, he is not currently counting hawks. This is another distinguishing aspect of hawkwatching—that of longevity. That is, are you able to consistently watch hawks year after year? The deck is stacked against most. Finding time, money, and support are always difficult. Family, health, commitments, and desire fluctuates over time. All these factors seem to collude against the hawkwatcher and hawkwatching. This is why having a working hawkwatch is so exciting; everyone knows how ephemeral it can be.

There is also a synergy when working with other hawkwatchers. Each member contributes a mix of specialties, thereby making the group stronger than the sum of the individuals. In Kevin Georg's case, he works with other experienced hawkwatchers. Erik Bruhnke is one such individual. He was the Count Interpreter at Hawk Ridge Bird Observatory in Duluth, Minnesota, so you know he can handle the cold. Erik is skilled at identification both near and far and a genuinely nice guy. I met Erik at The Biggest Week in American Birding (held each year in northwest Ohio during the first week in May) and sat through more than one of his hawk identification presentations. He was as charismatic as he was informative. Erik really made the art of identification fun for everyone, and I am sure he provided complementary skills to Kevin at Corpus Christi.

I, like so many others, have always wondered how hawkwatchers manage to watch hawks full time. The basic act of breathing each day requires a certain amount of money and resources. Hawkwatchers must eat, shower, sleep, and shop like other human beings. On top of the bare necessities, hawkwatchers need gas money, appropriate clothing, sunscreen, as well as equipment. So how do hardcore, full-time hawkwatchers survive?

There are a few strategies for existing as a hawkwatcher. The first is to simply become a birding "bum." The term bum is not meant in a derogatory sense, but implies an alternate lifestyle that is devoid of a typical nine-to-five job, one in which residency is fluid, and one that is opportunistic. Again, none of these are negative attributes; they are just different. The highly-respected Kenn Kaufman was kind of a birding bum as a young man, hitchhiking across the nation in an exploration of birds. Today, he is the author of an entire series of field guides and gives talks all around the country.

Birding bums go with the flow and accept offers from other birders who live vicariously through them. There is a romantic sense of being ostensibly one with nature and lacking any pretentiousness—taking what is given and making the best of it. Hawkwatchers can fit into this lifestyle just as easily as snowboarders or Grateful Dead Deadheads. The benefits of this lifestyle are living light, being mobile, and not having any possessions to worry about. There are no deadlines, no commitments, no bills, just living for the day with an optimism that tomorrow will be alright one way or another. The negatives of such a lifestyle are that hygiene may suffer, food may be scarce, and, if a problem does arise, you are at the mercy of other people's charity. If no one is around, then you are at the mercy of nature herself.

Another strategy is one of security. Some birders have worked early in their lives and have been able to build a sufficient repository of currency so that they are financially sound. They may have stocks, a pension, social security, and/or investments that keep them looking more towards the birds than towards the next workday. Many retirees live this style of hawkwatching. So long as they live within their means, they can spend as much time hawkwatching as they like.

A hybrid arrangement is sponsorship. In this type of lifestyle, hawkwatching becomes a means of sustenance. Arrangements are made for some remuneration for the services provided to the hawkwatcher, albeit serving as a hawk counter, a photographer, a researcher, or hawkwatching host. Sometimes, a salary is the means of economic transfer, but other times a blended model of payments and benefits are negotiated. Some counters receive stipends for counting during a season. This is usually monetarily insufficient for living; but, combined with a campground site, a room in a host's house and/or becoming a caretaker of a public residence can suffice.

In Kevin Georg's case, he has done it all. He has lived in his car, a camper, public housing, and even a rustic cabin. He has gone without! Nowadays, Kevin lives off his pension, gets a stipend from Hawk Watching International, and is able to live a stable life watching hawks seven months of the year. Kevin's lifestyle is a dream for some, but it is hard to say if they could live it. Essentially, Kevin watches hawks every day for months at a time. The isolation also affords few activities other than hawkwatching; so, even on his days off, he still shows up.

The type of hawkwatching that Kevin does in Corpus Christi is very different than what most hawkwatchers experience. When the birds begin to flow, it is like a river. Here, clickers are not used for individual birds; instead, they are clicked for every 10 birds. Some days, a click may need to mean 100 birds. This type of mentality may explain why Kevin sometimes blurts out an identification before he has positively made it.

I noticed this at the Allegheny Front, when Kevin was attempting to identify hawks that were still very far away. It was as if 10,000 birds were lining up behind this one bird, and that he had better get to it as early as possible. Of course, at the Allegheny Front, 10,000 birds would be a good number for the entire year.

Not everyone is convinced that Kevin has seen the most hawks. Not everyone thinks it is even possible to count that many hawks. These very human reactions are part of our psychology.

Humans evolved as eusocial creatures, which is the pinnacle of social systems. By socializing, we learned to cooperate, which enabled us to build a complex civilization. This required that individuals become reliant on others for living. In exchange for making shoes for your family, you needed to share the crops you grow in the field. Life was no longer dependent on how many different skills an individual had, but rather how specialized you were in an individual skill. To ensure that the farmer would share the crops after the shoemaker made all of the shoes, or, inversely, that the shoemaker delivered the shoes after being fed all year, some type of feedback system was needed. This was the role of religion and social norms, which functioned as laws well before jurisprudence (a.k.a. theory and practice of modern law).

As we evolved socially over the millennia, we became more and more socially and genetically programmed with emotions that helped with checks and balances of our behavior and the behavior of our neighbors. Today, children as young as toddlers have an innate revulsion to cheaters. The universality of hatred for thieves, liars, and cheats is ubiquitous in all of our social activities. We scrutinize what others say and lambaste those who break the social norms. We especially keep our eye out for those who have more than we have. A man who has a million dollars in his bank account is more suspicious than a man that has only a thousand. Does he deserve to have a million dollars? How did he get this money? Did he earn it honestly, or did he use deceit to make his fortune? Is this man a king?

I had heard muffled talk about Kevin Georg long before I met him. "How can you possibly put that you saw 83,994 Turkey Vultures on hawkcount.org? Are you sure it wasn't 83,993 Turkey Vultures? Did you really look at each bird, or did you just count that cloud of birds and that cloud of birds?" Other comments include, *"Who does he think he is? He must think he is special with all of his raptors! He is no better than us! I wonder if he just makes these numbers up for fame and fortune."* And even more, *"He thinks he is better than us, here at our rinky-dinky hawkwatch. He has no idea about us or this place."*

I suddenly hear a man speak, "Red-tailed just flew above us."

My mind suddenly clears as I recognize that I am counting at Allegheny Front Hawk Watch, and the man's voice is that of Kevin Georg. I was daydreaming for a moment. I snap out of it and realize that Kevin is on the picnic table next to me. He is leaning back and looking through his binoculars. No one else is close, and the muffled voices were all in my head.

Kevin Georg was here in the beginning days of Allegheny Front. He was one of the people who put this hawkwatch together. He has been counting here with that veterinarian he met years ago. That veterinarian, by the way, is none other than Tom Dick. Kevin Georg was instrumental in developing the protocols that are still used today.

Kevin puts down his binoculars for a moment and says, "Did you see that Red-tailed Hawk?"

I reply, "No!" I had been looking at the sky while my mind was wandering and did not see any bird.

"It was up high! You can probably still catch it." I run over, point my binoculars up and behind me. It is a Red-tailed Hawk, and it is really high.

I ask, "How did you see that one?"

Kevin, looked at me with a big smile and said, "You gotta scan, man!"

I replied, "I think you are right!" I had felt guilty about all of the silly thoughts that I had been thinking of just a few minutes earlier. I then added, "I am glad you are here, back home… King Georg."

He put his binoculars down, smiled and questioned, "King Georg?"

I said "Don't worry about it." Kevin started to scan again, as I then added under my breath, "King Georg."

A MAGICAL NOVEMBER DAY

It is November 21st, the tail-end of Golden Eagle season. Excitement is in the air. The Allegheny Front Hawk Watch holds the Golden Eagle record with 279 birds passing in a single season. That record, set in 2011, is the hawkwatch's shining moment.

Allegheny Front is a wonderful hawk site, but the total numbers of birds counted is not extraordinary. This is because the plateau that the hawkwatch sits upon is the eastern edge of the main migration route. Migrating birds further west are less numerous and more dispersed. If birds are migrating along these ridges and the winds are coming from the west, the birds tend to get pushed east, away from the Allegheny Front and out of view.

The Allegheny Front Hawk Watch is, therefore, wind-direction dependent, and it is the east wind that makes the great views possible. Unfortunately, air currents sweep across North America from the Pacific towards the Atlantic, thereby making west winds the norm. It is an unfortunate set of factors that works against the Allegheny Front, and yet it remains one of the most loved hawkwatches. It is an underdog, born with an impediment of needing a wind that is so scarce.

When an east wind is called for, hawkwatchers mobilize and head to the Allegheny Front. However, not all birds are as finicky about the wind direction. Birds that flap their wings more tend to be less influenced by the direction of the wind than those that soar. Golden Eagles prefer to soar and glide and, therefore, are less numerous on west wind days at Allegheny Front.

Today is a southeast wind, and the season total for Golden Eagles is 269—just 10 away from the record. I am counting here for just my fourth Saturday, and not everyone knows me or that I am now the Saturday counter. Hawkwatchers are clearly excited about the prospect of the day and begin rolling in at 7:00 a.m. Some are new faces, but I know who they are.

Serious hawkwatchers, especially counters, have a particular look. It is hard to describe, but it is obvious. They, long ago, gave up worrying about how they appear and dress purely functionally, mainly for battle with the elements. Full snowsuits, scarves, moon boots, hunting apparel, oversized gloves, and ski masks are regular features. If it works, they wear it, often in a mismatched fashion. Binoculars are usually held in place using a set of straps that look like a bra minus the cloth.

They also have that stare—one that scans deep into the sky like radar. These seasoned hawkwatchers tend to stand where the views are the best, which often coincides with where the wind is blowing. At 7:30 in the morning, these chaps are standing in 12 mph wind that is below freezing, staring in anticipation.

I introduce myself, and we recognize each other's names immediately, for hawkwatchers not only check their home site on hawkcount.org, but also look at the surrounding sites.

"Aren't you the guy that saw the Peregrines down at Cumberland Gap?" asks one of the men.

"That's me," I reply. I then ask, "You're a regular down at Washington Monument. I've seen your name many times." As more and more hawkwatchers pour into the parking area from various states, it becomes obvious that today is expected to be a good day.

Some hawkwatchers travel like the birds, or more accurately, with the birds. They do not have one particular home hawk site but visit those that have the best prospects on any given day. These are the individuals who bring news and stories not posted on hawkcount.org. Within minutes, everyone knows everyone, and news of what is happening at the other sites circulates throughout.

I tell everyone that Allegheny Front is not known as an early morning site. "Things do not get moving until at least around nine in the morning here."

Immediately, I get some resistance for this statement from counters at other sites, who posit, "You are missing birds starting that late."

I explain that they are correct in their thinking, but the landscape at Allegheny Front often inhibits the watchers from seeing those birds. The soaring raptors are either flying down below the tree line or on the western side of the hill.

I finally agree with the men and state, "The birds may be flying, but we rarely get anything early in the morning."

For the next hour, I think deeply about starting earlier. These hawkwatchers are some of the very best from very productive sites, and they have just told me that I am starting too late. I cannot be too confident in my response, for these guys have wisdom. My brain scans all of the early days I have counted at Cumberland Gap Hawk Watch, and I cannot think of a single day where I have seen any birds before 9:00 a.m. I then think about all of the early mornings I have spent here over the years and again have a hard time recalling early morning raptors. Despite all of my internal data, I am suddenly worried that I must somehow be wrong.

I walk over to Che Mincone to discuss this issue. Che and his lovely wife, Marian, have been here from the start, when the Allegheny Front Hawk Watch was just beginning in the early '90s. Che is a stickler for protocols, following the rules, and, most importantly, collecting and reporting accurate data. His mind is scientific in nature, and he can recall dates, times, and figures with little effort. Because of this, Che is a vault of historical data. He also has a small but potent repertoire of semi-dirty jokes. He agrees that we are missing birds early in the morning but also agrees that they are often not able to be seen. Everyone agrees that not many birds fly between 8:00 a.m. and 9:00 a.m. at Allegheny Front. That is evident from having had the conversation during this hour without a single bird flying.

As the morning progresses, the excitement is palpable. But as the hours begin to pass, few birds materialize. During the nine o'clock hour, the first bird of the day appears. It is a Bald Eagle, and that is it for that hour. During the ten o'clock hour, two Red-tailed Hawks and a Golden Eagle pass. Some of the hawkwatchers get anxious. I reassured them by reiterating that this site is not usually known as a good morning site. It seems as though the eagles are late flyers here, and this day is not going to be the exception.

I was feeling a bit of pressure, wondering if the other hawkwatchers were beginning to doubt me and maybe even their decision to drive to Allegheny Front. Since many of the newcomers had just met me and had never birded with me, they had no reason to have confidence in my claims, my skills, or integrity. Who was I to tell the elders about hawkwatching? Many had been watching hawks long before I ever picked up a pair of binoculars. Many of these people logged extraordinary hours at various sites. Maybe I should speak less, listen, and learn. I knew that I was being a bit hyperbolic with this thinking, but I wanted everyone to have a good day. Many had traveled hours to be here, and I wanted them to have a positive impression of the Allegheny Front and of me as a counter.

At the beginning of the eleven o'clock hour, an accipiter finally appeared. We had been waiting for some action, and finally, a flapping bird was coming through quickly. Unfortunately, I was filling in the data sheet with the weather conditions when the bird came by. I was late on the binoculars, but at first glance, it looked like a Sharpie. That's a safe bet because Sharp-shinned Hawks are the most numerous of the accipiters. Unfortunately, distinguishing the

Sharp-shinned from Copper's Hawk is one of the most difficult aspects of hawkwatching. Despite my lagging start, I focused on the bird and got a quick glimpse of him. And, it was a him. With accipiters, the males are smaller than the females. The size of a female Sharp-shinned Hawk is very similar to a male Cooper's Hawk. I had spotted the white, rounded terminal band of the tail and noticed the rather straight front edge of the reddish streaked wings.

I exclaimed, "Adult male Cooper's Hawk." Calling a Cooper's is always news, for the average ratio of Sharp-shinned to a Cooper's is about 10 Sharpies for every one Cooper's. By making this exclamation, I knew I would be scrutinized. But, there it was, a male adult Cooper's Hawk. A few of the veterans looked over and gave a reassuring "Good call." I was relieved.

Everyone in the hawkwatching community has an opinion concerning techniques of identification, protocols, and best practices. All thoughts are geared towards getting it right, and every hawkwatcher in the field knows how difficult it can be to make the proper identification. When I made that Cooper's Hawk call, everyone was with me—relieved that it was the right call; relieved that they agreed; relieved that I was one of them and not a charlatan.

In my mind, I thought, *"Let it begin."* And it did! Two Golden Eagles, a Red-tailed Hawk and an American Kestrel flew by, keeping everyone alert and ready. The one Golden Eagle remained unaged due to the lighting of the bird. Everyone discussed this eagle, looked at pictures, and compared verbal notes. We all agreed that we could not get an age of that Golden.

What is most interesting about hawkwatching is that the counter has the final say. In fact, it is improper to confer with the counter during the count. The first rule in hawk counting is that the counter's call is the only one that counts. Some would put it another way—the counter is always correct. The second rule is to reread the first rule. Because of this system, some counters sit quietly and mark their paper while the others discuss the bird. If they disagree with the counter, they are out of luck.

All hawkwatchers have witnessed bad calls. We have all made them. When it happens, it is usually let go without much fanfare. I recollect one bad call when a large bird came through a hawkwatch in early September. It was at eye-level with a dark bellyband and a bright red tail. It moved quickly, and the three of us hawkwatchers were happy to have caught the bird. Joe, the other hawkwatcher, yelled up to the counter, "Did you get that Red-tailed Hawk?"

The counter replied, "Yeah, it was a Broad-wing." It was not! Joe and I looked at one another for less than half a second and then began to rescan the sky. That is all the drama that you get for a bad call. There is no need for pleading your case. The compiler surely knows that you disagreed and will just as surely be extra careful with the next ID. If you are petty enough to push this issue, be prepared to also receive push back when you make a similar call. In the end, that one "Broad-winged Hawk" was one of only three birds for the day. The other two were a Red-tailed and a Broad-winged Hawk.

Again, the counters' code is that the counter's identification is the one that counts. If I would have put an age for the unaged eagle, it would have, because I was the counter, been marked as such. Nonetheless, counters are fallible, but they have at their discretion the use of other spotters and qualified observes to help make judgement calls. Some are comfortable using the observers as a resource for making final identifications—others are not.

Personally, I welcome the help, especially when aging raptors, which is most often done for eagles. The skill level is not the same for all eagles. Aging Golden Eagles is much tougher than Bald Eagles, for their plumage is more subtle from one age to another. The key is looking at the tail; but, when the Golden Eagles are gliding overhead, their tails are often tucked and obscured. Other indicators can be used, such as the pointiness of the individual flight feathers and the molting patterns. By detecting these cues, aging is possible.

Depending on the height of the bird and its velocity, it may be visible only for a fleeting moment. High quality pictures taken as the bird passes can be used as an aide. In real time, the image can be zoomed on the camera screen, yielding astonishingly nuanced aspects of the bird's shape, feather structure, and color. This process can increase the confidence in the aging process. Using this technique, I can confidently state that a bird is not only a sub-adult, but can often distinguish between a sub-adult III and sub-adult IV bird. Using cameras in this manner is revolutionizing hawkwatching.

The cameras were instrumental during the noon hour when an adult Golden Eagle appeared at 12:28 p.m. and a sub-adult at 12:41 p.m. Since the pace of the birds was a bit slower (just seven Red-tailed Hawks and one Red-shouldered Hawk), we spent time analyzing these two eagles.

The camera operators made fine adjustments to their cameras that optimized the lighting for contrast and brightness. When the one o'clock hour came around, we were ready for the four Goldens that came through. We were becoming an efficient working group! Everyone was gelling into a superorganism capable of identifying each bird in detail. It was like being in a high-level outdoor classroom, with experts filling in the very small gaps of the next expert. Everyone was simultaneously a student and a teacher. The learning was joyous and the participants jovial. It was turning out to be a most pleasurable day.

The one o'clock hour had 17 Red-tailed Hawks and Golden Eagles at 1:10 p.m., 1:31 p.m., and 1:40 p.m. We were just one away from matching the record. It had crept up so quickly, that the adult Golden that came in at 1:48 p.m. was the tying bird.

Without too much uproar, I stated, "We are tied for the record." The next bird would push us over the edge. Despite our excitement, we certainly did not want to overemphasize this record out of respect for our fellow hawkwatchers, some of whom were visiting from Waggoner's Gap, the hawk site that held the record before the Allegheny Front with 276 Goldens in 2006.

During the two o'clock hour, we were at a fever pitch, anticipating the record-setting Golden Eagle. The hawkwatchers from Waggoner's gap were excited and genuinely happy for us at Allegheny

Front. Unfortunately, the flow of Golden Eagles stopped. Excitement peaked when an eagle appeared at 2:06 p.m. It had some white on its wings. It flew right overtop of us, and it was low. Such a beautiful view, but it turned out to be white in the armpits. It is the first instance I recall hearing a disappointing howl, "It's just a Bald Eagle."

A Northern Harrier came through, which is a bird that typically creates excitement. Today was an eagle day, so it unfortunately did not receive the attention it was due. Another Bald Eagle, this one a sub-adult III, came through at 2:39 p.m. Aging this bird was fun, but everyone wanted a Golden Eagle. We wanted to be part of the record-breaking flight. Unfortunately, the Golden Eagles seemed done for the day—an early flight that had exhausted the population.

This thought seemed to pervade the group, but experience at the Allegheny Front demonstrated that late flights were the norm for Golden Eagles. I vanquished that thought from my mind, thinking that I may somehow jinx the count. For that brief moment, I seemed to have become superstitious. Clearly, I was caught up in the excitement.

The three o'clock hour was approaching, and the sky was losing its bright sunrays. Daylight savings had recently ended, so the darkness was earlier each night. *"We are running out of daylight,"* I thought. Maybe the Goldens were done for the day. Maybe I would miss the record. Today was the only chance, for the next day was calling for unfavorable northwest winds. I began to worry.

Then, at 2:48 p.m., without warning, the record-breaking Golden Eagle appeared and passed. I attempted to take a picture of the bird, but it moved too quickly. I wanted to ensure the bird's identification, so I stayed with my binoculars until the bird passed the centerline of the site; then, I jumped to my spotting scope.

My typical set-up is binoculars hanging on the left side of my body off a double sling harness and a D7000 semi-professional camera with a high-end 70-200 mm lens on the right side. I switched to the spotting scope because the light was becoming problematic, and the extra light grabbing scope allowed for a better view as the bird headed south.

Someone quietly muttered that that was the record bird, but the mood was surprisingly subdued. I am not sure what I had expected when the record was broken, but it turned out to be low-key.

Then, someone said, "Here comes another."

Everyone excitedly focused on this bird. "It's a Golden!" a watcher exclaimed. Suddenly everyone was very excited.

I am not sure why everyone was more enthusiastic about this bird. Maybe it was because the record bird was special to Allegheny Front more than to our current group that included visitors from all over. This 2:58 p.m. Golden Eagle was everyone's bird, and the joy was more inclusive. I felt a bit guilty about putting so much emphasis on the previous eagle and also became more excited about our current bird. This was *our* bird, and I was fortunate enough to have been given the chance to count it.

Did I deserve to be the counter on this day? Had I earned it? Clearly, I owed a debt to all of those who had stood on this special spot and counted all the other eagles in previous years. Especially to those who stood here when the eagle numbers were low, when the excitement was absent, when the weather was unforgiving.

People like Gene and Nancy Flament, who would drive from Pittsburgh to take their turn bearing the poor weather and who had made me feel so welcome as a new hawkwatcher years ago. Gene used to say in his old-timer drawl, "You'll get it, boy; give it time," when I would confide in him that I just could not seem to identify the hawks. He would let me stand near him and take a guess at the bird before giving me the answer. His use of "boy" was not calling me a boy, but just his way of connecting sentences. Gene has a soothing style of speaking and is a comforting soul. I was so glad he was here for this record day, and I thanked him.

I am not sure he knows how instrumental he was in his encouragement of my learning hawks. Other members, like Tim Anderson, were vital to my development. Tim was the Saturday counter and was one of the best hawkwatchers I have ever met. He spent many Saturdays helping me develop into a hawkwatcher.

Photo by Jeanine Ging

This picture was taken after breaking the Golden Eagle record at Allegheny Front Hawk Watch. From left to right: Gene, Randy, and Nancy Flament, Bob Stewart, an unknown visitor, Jim Rocco in front of Kevin Georg, an unnamed photographer that frequently visits the site, Brian M. Wargo, Janet and Tom Kuehl, Ed Gowarty, Paul Fritz, Joe Sabo, Joe Kelly, another unknown visitor, Che and Marian Mincone, Mike Smith

The three o'clock hour began with a slight break from the birds. For all of 17 minutes, we counted only a few Red-tailed Hawks. Then, a pair of eagles materialized—an adult Golden Eagle followed a minute later by a sub-adult II Bald Eagle. It was an unlikely pair, but today seemed like an unusual day where anything could happen. The Golden Eagle seemed special for some reason, like it had the largest crop of all time.

We yelled to one another, "Look at the crop! It looks like it just ate a rabbit." Maybe it had, but there was something unusual about

162

this bird. I examined the picture I had snapped and was "blowing it up" on the small screen of the camera. Then, I saw it.

There was a small transmitter attached to the back of the bird. You could see it when the bird flapped downward. Others also had pictures showing the transmitter. The metal-looking box on the back of the bird seemed to exacerbate the size of the bird's neck. Years earlier, the Allegheny Front collaborated with a firm that caught and placed the transmitters on the Golden Eagles.

Someone yelled, "I bet that was one of Mike's birds." They were referring to Mike Lanzone, who was responsible for fitting the birds with the transmitters. I made a note to myself to try to track this bird, but there were more pressing matters at hand.

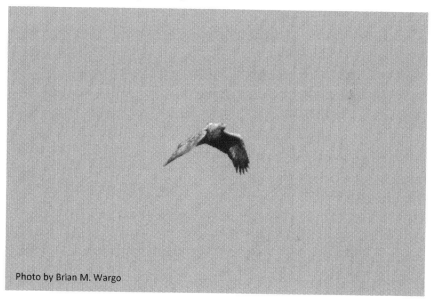

Photo by Brian M. Wargo

This adult Golden Eagle has a full crop and a transmitter on its back.

An impatient juvenile Golden Eagle gave us only six minutes to view the cropped Golden before stealing the show. Its bulging wings had the most beautiful white wing marks. This was, by far, the most picturesque bird of the day. Unfortunately, the sun was getting low in the sky, so the pictures did not display the full beauty of the bird.

Nevertheless, the image of this bird is burned into my brain. It is now my essential juvenile Golden Eagle.

The sky quieted for about 15 minutes and then unleased a sudden barrage of eagles. A sub-adult III Bald Eagle was followed immediately by an adult Golden Eagle. As I finished recording these two eagles, another adult Golden Eagle appeared. As soon as I was done with this bird, another took its place. And then another. And then another.

It was time to take the weather for the four o'clock hour, but I just recorded a 3:59 p.m. eagle, a 4:00 p.m. eagle, and a 4:01 p.m. eagle—all Goldens.

Someone jeered over to me, "How are you keeping up?"

I replied, "Not well!"

Everyone laughed and someone stated, "It is a good problem to have!" And they were right!

After finally recording the weather conditions, I figured that the eagles were just about done and that I had witnessed the grand finale. I took a breather, thinking, *"Well that was one hell of a good day."* I had good reason to think so. It was getting late, and the sun was going away. From all normal indications, the day was just about over, including the flight of the migrators. The thermals were gone; however, the 20 mph southeast wind remained.

Then, Kevin Georg made a proclamation at 4:08 p.m., "There is a bird out between the notch." Kevin is legendary for spotting birds that are merely dots in the sky.

I looked up and saw nothing. "I see nothing!" I replied. Then, it came into view—a dot with another dot behind it and another dot behind that. This was just the beginning of an unbelievable four o'clock hour.

Two eagles appeared almost side by side. A sub-adult Golden and an adult. This was immediately followed by another adult that was low and right in front of the hillside. It was so close, you feel like you could run to the ledge, jump, and grab onto the eagle's legs. The hawkwatchers not used to the views at Allegheny Front were simply flabbergasted by the close proximity of these majestic migrators.

It was just at this moment that another bird flew right behind everyone, and it was only about 100 feet above our heads. Kevin and I saw it, but no else did because they were focused on the Golden right in front of them.

I yelled, "Everyone see that Bald Eagle?"

Everyone turned and looked at me and said, "That was a Golden Eagle."

I said, "No, the Bald Eagle that almost hit the trees above us." No one knew what I was talking about. "There was a sub-adult IV eagle that just flew over."

They responded, "How could you tell if that was a sub IV or an adult in this light?"

I said "'Cause it was a Bald Eagle that just flew over." Everyone thought I had lost my mind.

Kevin Georg chimed in, "You guys missed this one! It was right here! You were looking in front of you, and only Brian and I were back here to see it."

Two members then asked me to come over and identify what they had caught in their camera. I obliged.

"What do you see here?" they asked.

I responded, "An adult Golden."

"That's what we are trying to tell you!"

I said, "I got that bird, but there was another directly overhead."

"This is it!" they insisted.

"Then where are the tree tops in your picture?" I asked.

They said, "This is the third bird!"

Kevin jumped in, "That *is* the third bird. We are talking about the *fourth* bird."

"What fourth bird?" they queried.

The conversation ended when the next eagle came through, which was just three minutes after the sub-adult IV Bald Eagle. It was 4:14 p.m., it was cold, and the wind was howling. I had been standing in the elements for nine hours, and I was drained. I counted the adult Golden Eagle, and everyone thought that must be the last one. The birds were getting lower and lower. They were clearly coming down for the night. It was almost dark at this point, and the sun was no longer offering any warmth. The last rays were shining through the trees. What a day! Some people began to pack-up and leave. They had witnessed a remarkable flight.

Then, at 4:24 p.m., an adult Golden Eagle came through, followed by another adult at 4:25 p.m. Everyone was laughing, thinking that it was over when it clearly was not. Then, at 4:29 p.m. and 4:37 p.m., adult Golden Eagles came through, each bird receiving more and more shouts of disbelief. Finally, at 4:54 p.m., the last Golden Eagle came through. It was just about five o'clock, and the count was clearly done for the day. Everyone was celebrating a wonderful day. Some people left promptly, having stayed longer than they expected, and were now running late. Everyone was excited about the day and ready to get some food and into warm cars.

I let everyone know that I was remaining at the site for a bit longer and that they were welcome to stay. My general plan was to begin the process of entering the data into hawkcount.org, update the Twitter account, and post something to the Facebook page, thereby sharing the day's events with those who were not able to partake.

The moon was already dominating the sky and was becoming brighter with each minute. The air was cool, the crowd excited, and the feeling was of pure joy for the marvelous day. People were just hanging out in the field, chatting and laughing. The long distance visitors were commenting on how nice of a time they had and how pleased they were with their trip. They repeated that they were blown away by the close views of the eagles, the inclusiveness of the group, and the absolute beauty of the plateau.

I was walking around thanking each and every person for their help in making the day a success, shaking hands and swapping stories. Then, Mike Smith, a regular at Allegheny Front, stopped talking and eerily stared into the northern sky. He stood silent, which

166

was odd during this celebration. I saw him out of the corner of my eye and was wondering what he was doing. He was like a dog hearing something outside and standing at attention. It actually freaked me out a little—too many movies exist where the big event is about to happen and only one man is chosen to know it.

I stopped and watched Mike, not the sky or what he was staring at, which was clearly nothing because the sun was down and the sky almost dark. I thought to myself, *"What is he doing? Is he listening for something?"*

Mike quickly raised his binoculars and loudly and sternly said, "We got one coming in!"

"What?" someone exclaimed.

Someone else said, "You have got to be kidding!" Everyone looked up and witnessed the bird sailing like a low-flying blimp. It was a Golden Eagle, slowly and silently passing overhead. The eagle moved out and away from the edge.

"It's going to go through the moon! It's almost there at the moon! Get it! Get it!" someone implored to those with cameras. I had my camera, but I needed to identify the bird.

Through the binoculars, I quickly identified it as an adult. The bird passed, and everyone was hooting and hollering like a bunch of kids at a parade. Here we were, standing in a cold, nearly-dark field in the middle of nowhere, enjoying a spectacle that most would not believe unless they were there. Thankfully, we were there! Thankfully, Mike was looking up. It was 5:12 p.m., and that was the exclamation point on the phrase, "What a way to end the day!"

The day was clearly over, and it was time to pack up; however, no one was moving. Within a few minutes, the very faint dots of the brightest stars began to appear in the sky, and everyone was enjoying the ambience. We had been rewarded with a picture-perfect ending— an eagle traversing the moon. It felt like we were characters in a movie. Actually, it felt more like a concert when the band walked off stage and the crowd was willing them to give an encore performance. Just like at a concert, the energy was building, not waning. But, for

167

what? Performers can be pressured into coming back out. Eagles cannot! At least that is what I had thought!

The stars were out, the wind had slowed, and the far-off lights from distant towns were lighting up the landscape down below the plateau. I was tired, hungry, cold, and yet I was not ready to leave. No one wanted the show to end! Then, Mike raised his glasses again, this time joking, "Here comes another!" We all laughed, but Mike was not laughing. Right above us, just over the tree line was a silhouette of a Golden Eagle. Everyone just stared dumbfounded into the backdrop of stars.

"My God, they are still flying!" stated one of the Joe's.

"He's got a friend!" someone yelled.

"*What?*" I thought.

Clearly, they had to be wrong. And they were! There were *two* friends joining the first, not one. Everyone was ecstatic and filled with emotion. I remember looking over at Paul Fritz, a respected hawkwatcher who stoically stares into the sky for hours at a time without changing the expression on his face. This is a guy who has done some serious hawkwatching. The smile on Paul's face is what I remember even more than the eagle's silhouettes. He looked like a kid who had just been given a new bike.

I could not see Joe Kelly, but I could hear his infectious laughter. Joe has been around the world and birded everywhere. But tonight, I think Joe would have chosen this spot over any other. I was so happy when Paul and Joe both excitedly stated, "I will definitely be back! I am now a believer in Allegheny Front!" and "I will remember this day for the rest of my life."

Everyone seemed to know that we could never top seeing eagles in the starry background. The party was over, and I recorded the last three eagles as Goldens, marked at 5:22 p.m., 5:23 p.m., and 5:24 p.m. Everyone shook hands, patted each other on the back, got in their cars, and drove off.

We shared something special that night, and everyone knew it. As the taillights left the parking area of the Allegheny Front, I stayed and added the total birds for the day. Ninety-three raptors for the entire 10-hour day. The breakdown was a Norther Harrier, a Cooper's Hawk, a Red-shouldered Hawk, an American Kestrel, 51 Red-tailed Hawks, six Bald Eagles, and 32 Golden Eagles.

These numbers are a pittance at many other sites, averaging fewer than 10 birds per hour. That is where the numbers only tell part of the story. Unlike many other big number hawk watches, Allegheny Front is special because...well...because...because...it's...it is hard to explain!

Yes, the views of the birds are unusually close. Yes, the people are very welcoming. Yes, the site is beautiful. However, this type of reductionism just doesn't work. It's just an enchanted place—a location in space and time where people gravitate, rejuvenate, and remunerate their inner being. It is difficult to put into words, and more so in describing this particular day. As I uploaded the report, I wrote the following in the non-raptor notes:

A truly magical day with 32 Golden Eagles! The first GE was at 10:57 a.m. and the last three were spotted as the stars were visible in the sky. The Eastern Flyway record for Golden Eagles in a season, held by Allegheny Front Hawk Watch since 2011 at 279, was broken with the 10th Golden of the day (2:48 p.m.). The new record is now 302. It is hard to describe the excitement, comradery, and sheer joy this day brought to so many hawkwatchers. Special thanks to the Waggoner's Gap crew and the veteran hawkwatchers at Allegheny Front for the great spots, pictures, and identification skills they brought to the group. This was hawkwatching at its very best and it was a privilege to count hawks with such a wonderful group.

It was the best I could muster from my cold fingers and exhausted brain at the time. Even now, I cannot do any better. Some things are impossible to capture with words, especially when dealing with magical places like the Allegheny Front Hawk Watch.

ESSAY TWENTY-ONE
HAWKWATCHERS NOT WATCHING HAWKS

I t is a strange and unnatural meeting of hawkwatchers. No one is wearing binoculars, no one is dressed for the elements, and no one is looking up. Instead, this group of hawkwatchers is sitting around a gigantic table talking about budgets, record-keeping, and discussing educational initiatives.

It is mid-November, and I am at the Acopian Center for Conservation Learning at Hawk Mountain Sanctuary. I am attending a board meeting of the Hawk Migration Association of North America (HMANA). For the next eight hours, everyone will be focused on the big picture of hawkwatching—who is doing it, how they are doing it, what needs they have, which new projects are in the works, which new policies, regulations, and technologies are becoming realized, and how HMANA can best serve to ensure that hawkwatching remains a viable citizen scientist activity.

For those new to hawkwatching, HMANA is a non-profit organization whose mission is to support the collecting, managing, and disseminating of hawk migration data. It was formed years ago when birders, field scientists, conservationists, and banders finally got organized and developed protocols for collecting and distributing hawk data.

Communication in the early days was somewhat clumsy. Think physically submitting sheets via snail mail, using dial-up networks, and worrying about rates for long-distance phone calls. Today, modern technology provides smooth and refined access of information, tools, and other hawkwatches.

Sitting in the Acopian conference room gives one a sense of the history of hawkwatching. Bookshelves are lined with journals, books, and binders filled with data. Interspersed are trinkets, pictures, figurines, and paintings of hawks. Archaic devices for collecting hawk data are also scattered around the room.

These artifacts are juxtaposed with the biotic instruments positioned at the center of the room. Here, the human capital, in the form of dedicated hawkwatchers, provides the impetus for the successful organization of otherwise disparate information and skills. Each hawkwatcher has driven or flown in from around the country to provide their professional expertise to the organization. For most, they are doubly volunteering their time—for they also devote hours at their home hawk sites. Today is just another day in the life of a hawkwatcher, and, thankfully, these diehards are up for the challenge.

This particular board meeting is open to all HMANA members; so, I, along with fellow hawkwatcher, Kevin Georg, decide to make the long drive. We meet at the Allegheny Front at 4:15 a.m. and make the three-and-a-half-hour pilgrimage to Hawk Mountain. Snowflakes are falling when we begin the journey, and the air is blustery cold. But, we are in a heated metal and glass box traveling down the road. A perfect time to do some bench hawkwatching. We compare stories, tell each other about our best and worst days, and discuss the future of our hawk sites.

The time passes quickly, and we arrive at Hawk Mountain. Upon entering the Acopian Center, we recognize other hawkwatchers either by sight or by name. *Hawk Migration Studies*, HMANA's biannual journal, is distributed to all HMANA members and reads as a "who's who" in the hawkwatching community. Reading any issue gives the pulse of the hawkwatching endeavor, and the pictures connect names with faces. I have read all of the issues from the last few years, so I feel acquainted with many of the members.

As Kevin and I walk into the conference room, I recognize the board members, each of whom warmly greets us as if they also recognize us, which they most likely don't. The introductions reveal that almost everyone is connected to everyone else in some way. Within minutes, I feel at home and settle into a chair for the morning.

The meeting starts abruptly, and the board jumps into the day's agenda. At the helm is Carolyn Hoffman, who chairs the board. Carolyn is a friendly face, but holds a heavy gavel at the table. She is a stickler for keeping on task and on schedule. She is just what is needed to keep such a large group on task.

As the meeting proceeds, Carolyn remains open to new ideas and encourages creative solutions, all while keeping an eye on the clock. Each member speaks professionally and deliberately with very little small talk. Everyone seems to sense how quickly the day is about to unfold. Time is precious here, so every discussion is on point.

Board members are constantly communicating with one another during the year, but board meetings are special because everyone is physically present. This requires elected members to travel, which is an arduous task any time, but is made more hectic because hawkcounting is continuing as they speak.

Everyone is tired from the long season but remain hyped about the birds that are still streaming through. Those thoughts and emotions must temporarily be suspended so that ideas that have been churning for months can finally be addressed by the board.

Every item that is presented at this meeting is thoroughly discussed, with each member speaking clearly and efficiently. Several committees report on their progress since the last meeting. Carolyn pushes each committee for clarification. She clearly runs a tight ship and is respected by all the board members.

After an hour or two, Carolyn breaks members into committees, each with a particular task. Without hesitation, everyone separates into groups and begin deliberations. Committee members brainstorm a variety of notions, but categorize each as achievable or fantasy.

Applicability and achievability are paramount in these discussions. Everyone knows that ideas are cheap, but enactment requires copious effort; so, only those ideas that are realistic are given attention. Unlike boards of major corporations, there will be no handing off ideas to subordinates for implementation. The committee members are the individuals that will need to enact these plans.

The committees burn through their hour or so of work in what seems like a very short period of time. The entire board meets back at the main table to discuss the major points from each breakout session. While all organizations form committees in an effort to seed creativity, committee work is often thought of as purely theoretical, rarely leading to anything substantive. Often, the only tangible work a committee produces is setting up another meeting or in the formation of another committee.

This is clearly not the case if you are sitting at a table chaired by Carolyn Hoffman—you better have specifics at hand. She will ask you not only for a summary of the committee's discussion, but will require that you set definite dates for accomplishing particular tasks. If you try to wiggle out by claiming that the idea is just in its infancy, she will want a date at which it will be mature enough for clarification. It is best to give her a date, because, if not, she will suggest one for you.

Everyone responds well to Carolyn's demands, for they share her passion and desire for results. The day progresses quickly and runs like a well-oiled machine. Every minute of every hour is focused and aligned to the agenda. Interspersed throughout the meeting are presentations from scientists, economists, and conservationists.

Despite the tight schedule, Carolyn and the board make time for general comments from HMANA members who attend the meeting and are sitting on the periphery. This inclusiveness and attention to members' needs is an exemplar of how organizations should run. Everyone is involved, everyone is focused, and everyone is amicable.

Time flies by, and before anyone can comprehend it, the meeting is over. Many will attend the dinner social that is scheduled for 5:00 p.m. and will continue talking. Everyone is welcome to attend. Kevin and I decide that we should get back on the road and, therefore, decline the invitation. As we exit the conference room, almost every board member thanks us for our attendance, gives their contact information, and invites us back.

Driving back west towards the Allegheny Front, I pondered how eight hours could progress so quickly. Probably because everyone was so enamored with the progress this organization has made since its inception and the promise that it shows for the future. By any account, the day was successful, and everyone left with assignments that were to be completed before the next meeting.

Attending the HMANA meeting exposed me to a small army of individuals that dedicate so much to the hawkwatching community. It is this unseen side of hawkwatching that is just as important as the counters and the hawk sites. Despite the lack of binoculars and spotting scopes, everyone was focused on the hawks as intently as if they were out in the field. This is a different kind of effort—one that most hawkwatchers are unaware of. Keeping the association running, paying the bills, supporting hawk sites, and getting the publications out on time is a lot of work, and Carolyn Hoffman and the entire board should be commended for their efforts.

Thinking about the meeting, I was impressed by the level of support that everyone offered to the association and to one another. Every idea, change, or query requires human capital, and someone always selflessly volunteered for the task. Usually, another member would join in to assist the original volunteer, thereby spreading the work into more manageable chunks. How odd it is that individuals create work for themselves knowing that most of their efforts will go unrecognized. This seems to be the mantra of hawkwatchers: Work hard in silence, without glory or pay, because of the hawks…for the hawks.

At first glance, it may seem equally odd writing about hawkwatchers that are not watching hawks; but, it should be pointed out that many people live vicariously through hawkwatchers that do. The hawkcount.org website is an apt example, for it allows those who do not or cannot watch hawks to watch hawk numbers.

Many hawkwatchers are not physically able to watch hawks because they are geographically isolated from migration hotspots. Others used to watch hawks, but their bodies or eyes are no longer capable. Most simply have jobs, families, and commitments that preclude them from devoting time to actual hawkwatching. Hawkcount.org allows everyone to keep abreast to what transpires each day at each site. They may still partake in the migration even though they are not physically present. In mind and spirit, they are there!

I personally check hawkcount.org every day to see bird numbers from both distant and near sites. I feel especially connected to the sites I have visited. I image the people, their conversations, and what their day is like. I also share in their excitement or disappointment as they post their numbers. I know they do the same when I post my numbers.

It should be noted that not all hawkwatches post their data to hawkcount.org. Wind turbine companies hire hawkwatchers for environmental studies, and the data that they pay for is considered proprietary knowledge. This practice rubs some people the wrong way, but they too are hawkwatchers. Their motives are just different. Those of us that do it for free or freely post our data do so for the entire community, both scientific and our social networks.

It is heartening to see so many work so hard to keep hawkwatching, a mostly volunteer activity, up and running. But we do it for ourselves as much as we do it for the hawks. It is part of who we are. We are hawkwatchers all of the time, even when we are not watching hawks.

I should probably confess that I am a hawkwatcher that spends copious hours not watching hawks. That's because I volunteer to write the Eastern Flyway Reports for *Hawk Migration Studies*. These analyses are a staple of the publication, summarizing the fall and the spring migrations. This big picture analysis helps provide perspective about entire seasons from great swaths of land from around the country. Comparing seasons gives a general state of the hawks and either gives us pause, hope, or comfort.

Writing flyway reports is hard work, consumes a significant amount of time, and is always subject to critique. My eyes were wide open applying for the position, and yet I felt compelled to volunteer. Like everyone else at the HMANA meeting, I felt I was contributing to the hawk community using my particular set of skills. This work needs to be done, and someone must do it. So, why not me!

Hawkwatchers that do not watch hawks are as valuable as those that do. They deserve our respect for the thankless work they complete. Without their efforts, our efforts might be in vain. I personally think of the HMANA board members and the work they do, the support they give, and their continuing drive to improve the hawkwatching endeavor when I am out counting hawks. They are with me, and I am with them.

Together, as a community, we participate as a single superorganism, trying to find meaning—scientific and otherwise—about the hawks, about each other, about ourselves. This example of self-organization is one of the great wonders about humanity and nature. It seems to defy our notions of entropy and randomness.

Maybe hawkwatching is just a glimmer of our capacity to connect ourselves with the biosphere. Maybe hawkwatching is just a comic metaphor for life itself—a self-referential, self-signaling, self-sustaining energy flow. Or, maybe I just need Carolyn Hoffman to break my delusional daydreaming and ask, "Brian, do you have that analysis for the fall reports done yet?"

I would reply, "No ma'am, Mrs. Hoffman! I will get to them right away!"

ESSAY TWENTY-TWO

THE FINAL COUNT

It is the end of hawkwatching season. The cold temperatures, the limited hours of sunlight, and the dormancy of life are signs that it is over. Everything appears dead except the wind. It tries to breathe life back into the brown landscape, incessantly blowing the trees, shrubs, and grass from side to side. The snow crystals make the usually invisible wind visible, showing its turbulent nature.

December usually offers finality to the migration, but this is an odd season. El Niño has struck with a vengeance, warming the atmosphere to unheard of levels. This December has been the warmest on record, and it is just now showing its typical attributes. The wind is gusting from the west at 40 mph, and it is far too blustery to stand outside. I am sitting in the car, but the wind is causing it to rock back and forth. The sun is trying to illuminate the sky, but this squall is hampering its efforts. This cold snap might encourage the late flyers to move south.

The official count ended days ago, but hawkcount.org reveals that Bald Eagles are finally moving. The last few days, several hawk sites reported Balds by the half dozen. A few Golden Eagles were also mixed in, but it is hard to know what is happening because only a handful of sites are still reporting data.

Despite the unfavorable wind, there is a chance that some birds may be seen today. This December has provided only eight birds for the entire month—the lowest in the history of the Allegheny Front. After such an extraordinary October and November, the Golden Eagle migration has shut down. On average, December produces 15 of these hearty birds; but, this year has only yielded one. What is causing this abnormality is anyone's guess. And, we are all guessing.

One idea is that the birds have no signal to leave their northern habitat. The weather has been warm, there is no snow to inhibit hunting, and it feels more like fall than winter. So, why bother migrating? Those birds that decided to move may have found a comfortable rest stop along the way and decided to simply stay put rather than carry on. The hunting season has littered the landscape with carrion, which is like having a free catered dinner. The carcasses are already opened up, making access all the easier. It is a scavenger's dream!

Maybe climate change is affecting the eagles? However, if that were true, why was there a surge in October? Shouldn't the birds be migrating later? No one seems understand what is happening. Allegheny Front is not just an anomaly; other sites are reporting similar outcomes. That is why I am out here. I am curious to see if the eagles are just waiting for a reason to migrate. If they do move through and no one is counting, then we are left in the dark.

Regardless of the outcome, it will be interesting; although, it may also be problematic. If the birds come now, we do not have anything to compare it to. The count used to end November 31st, but that was moved to December 15th after finding that the eagles were still moving in December. Now, if we find that the eagles are moving later than the 15th, what will we be able to say? Since we have not counted this long in the past, we will not be able to make the claim that this year is an abnormality. However, if they do not come, then we truly have an odd season.

We often think that we understand what is going on with the birds and with nature, but we may just be fools. Our pattern seeking brains love to assign causality to events. Thankfully, we have statistics that can help us demonstrate if a correlation exists. But, even then, we must be very careful because correlation does indicate causation.

We view the world through anthropomorphic eyes. We assume that the world that we perceive is identical to that of the eagle. In reality, we have very little evidence that this correspondence is valid. We assume that the cues that we receive are processed similarly in other species.

Sight is something that we clearly share with our avian friends, but to what extent their decisions and actions are dictated by that information is not as clear. They may prioritize their sensory input differently. They may even have senses that we do not. Humans, for example, are not able to utilize the electric and magnetic fields in any noticeable fashion. It may be that the birds can. This is just one example, and there are a plethora of potential sensory stimuli that we are unaware of because we are blind to them.

What I am certain of is that the day would be more enjoyable if something was flying in the sky. Earlier in the day, a few crows were performing acrobatic maneuvers in the gusty wind, but they moved on quickly. Since then, only the movement of the car during the gusts has provided any type of stimulation. And yet, I am reasonably content.

It is surprisingly warm and comfortable sitting in this enclosed space. The windshield, windows, and sunroof provide a view of the sky with surprisingly little obstruction. The sunlight heats the car via the greenhouse effect; so, despite the -13 degree Celsius (8° F) wind chill outside, I am able to circumvent the use of a coat, thereby making me more nimble and free to move around my seat for better views.

The silence inside the car is also rather astounding. Despite the howling wind, it is remarkably peaceful. I feel fortunate to be human, with the ability to conquer the elements. We have the wherewithal to engineer a movable bird observation station. The seats are padded, ergonomically correct, and adjustable. When I am done for the day, I will be able to turn a key, magically awaken this horseless carriage, and drive back to my home. Watching hawks in an earlier period would have been much less comfortable.

A single raven appears just over the tree line and quickly vanishes. Aloud, I state, "raven." Immediately the windshield fogs over. I make a note to myself that there is no need to explicate which bird is in the sky because I am sitting alone, and doing so actually inhibits the view. I open the door and exchange some air with the outside. It is then that I notice that my visibility is about to change from 15 km to less than one kilometer as a quick squall passes.

Within 10 minutes, visibility returns. This weather is treacherous, with large Cumulous clouds racing across the sky. A plane passes directly overhead, moving opposite to the clouds, and the combination looks like the plane is going Mach 5. I think to myself, *"If I were an eagle, I would surely fly south. Today! Right now!"* I scan the sky and, again, it is devoid of avian life.

This is the latest that I have ever watched hawks. In previous seasons, I was done at the end of November. Being at a site in which you can observe from the car is a luxury. Everything is infinitely easier. Eating lunch without worrying about your sandwich or its wrapping blowing away is fantastic. Having feeling in your fingers and not worrying about frostbite if you take off your gloves is even better. All of these conveniences illuminate how unpleasant it can be hawkwatching in poor weather.

The clouds cover the sun for a few minutes, and the temperature in the car drops precipitously. Without the sun's rays, the vehicle's energy input decreases. Suddenly, the day feels colder. Thankfully, the rays reappear, and warmth again spreads throughout the car.

From inside this controlled environment, it is a beautiful day. Windy, sunny, but brutally cold. Again, the raven appears, flying over the treetops, seemingly oblivious to frigid temperatures. It is as if he is taunting me to come outside. This act of defiance demonstrates that only the tough belong out here. The raven's message is received, and I suddenly feel small, helpless, and vulnerable.

As a species, we cannot survive without our technology. We lost the hair on our bodies, our ability to drink pond water, and any real defense against predators. All we have is our intellect and our wonderful hands that enable us to create, modify, and record. Out here, the raven wins. Thankfully, I am in here—safe from the elements, safe from nature's wrath.

The sky transitions from bright to grey to bright again. The wind follows this rhythmic path of blowing, then gusting, then just blowing again. Occasionally, the dusting of snow creates small vortices that resembles a tornado. It dies as quickly as it is born. This is a harsh landscape, and it is a wonder that anything is able to grow or live here. Nevertheless, life clearly goes on here, even in the winter.

The Corvids are clearly here as well as other creatures. Just a week earlier, Bob Stewart and I found this first hand. We were cleaning out the bluebird nest boxes, and we discovered a whole nest of life. After removing one side of the box, Bob reached in to pull out all of the nesting material and inadvertently grabbed a handful of mice. For a brief second, he held six mice in his hand. They jumped, hit the ground, and, within another second, were securely under a large rock.

Bob was stoic during the entire situation, never jumping, squealing, or complaining. It was simultaneously scary and funny. Most people would have freaked out, but Bob just looked up and said, "I guess there is still something living in there." It was hilarious and unnerving. Undeterred, Bob cleaned the other boxes. I recorded him with my phone, hoping to recreate the original experience. The other boxes were uninhabited, and we made a note to mouse-proof the boxes in the spring.

The hours pass here, and my hope of seeing eagles today are fading. Reality beckons, and I accept that today will mimic the previous days in December, being devoid of eagles. It is just about the shortest day of the year, so I know that the sun will retire early. My feet are beginning to get cold, so I put on boots while sitting inside the car. I then decide to put on a coat and a hat. The sun's meager light no longer heats the car. I consider running the engine, but decide against it from an environmental standpoint.

I am suddenly aware that I am cold, especially my feet. The one drawback about being in the car is that you sit the entire time. Without walking around or even standing, the body generates little heat. I contemplate walking around a little, but that is not appealing, and it may cause further cooling of my body. I decide that is simply too cold and stay inside the car.

I noticed some crows flying around and follow them. They like to greet and then harass other birds. Unfortunately, no other birds are here to harass. As I stare into the void, I recognize how much the birds add to the ambience of the atmosphere. Without them, the sky remains pretty, but less appealing. I like looking at the sky; but, now I recognize that, when I look up, I really like the sky, because it is where my feathered friends are.

I scan the blue with my binoculars and notice white, fuzzy streaks. I focus the binoculars and recognize they are small snowflakes moving very rapidly across the sky. A Common Loon moves through the field of view. I watch it for a few moments as it disappears into the white haze. I also notice about two dozen Tundra Swans getting blown around. They are far out and rather high in the sky. They soon disappear like the loon.

It is about 3:30 p.m., and I am cold. I have been sitting here for six-and-a-half hours, and the round-trip drive adds another four hours. It will be another two hours of sitting on the drive back home. I call it a day. I am indifferent about the day's results. Maybe the eagles are done for the year. Maybe I am also. I pack up and start the car.

As I drive home, I blast the heat. It takes about an hour for my feet to warm. It has been a very long season. I have spent every non-workday watching hawks since the middle of August. I am tired, and, as I head home, I feel like I am migrating away from here for the season. But, like the hawks, I will be back next season.

Made in the USA
San Bernardino, CA
22 June 2016